SCHOLAR Study Guide

Higher Biology

Authored by:
Bryony Clutton (North Berwick High School)

Reviewed by:
Fiona Stewart (Perth Grammar School)

Previously authored by:
Eileen Humphrey
Fergus Forsyth
Jaquie Burt
Lorraine Knight
Nadine Randle
Patrick Hartie

Heriot-Watt University
Edinburgh EH14 4AS, United Kingdom.

First published 2018 by Heriot-Watt University.

This edition published in 2018 by Heriot-Watt University SCHOLAR.

Copyright © 2018 SCHOLAR Forum.

Members of the SCHOLAR Forum may reproduce this publication in whole or in part for educational purposes within their establishment providing that no profit accrues at any stage, Any other use of the materials is governed by the general copyright statement that follows.

All rights reserved. No part of this publication may be reproduced, stored in a retrieval system or transmitted in any form or by any means, without written permission from the publisher.

Heriot-Watt University accepts no responsibility or liability whatsoever with regard to the information contained in this study guide.

Distributed by the SCHOLAR Forum.

SCHOLAR Study Guide Higher Biology

Higher Biology Course Code: C807 76

ISBN 978-1-911057-36-9

Print Production and Fulfilment in UK by Print Trail www.printtrail.com

Acknowledgements

Thanks are due to the members of Heriot-Watt University's SCHOLAR team who planned and created these materials, and to the many colleagues who reviewed the content.

We would like to acknowledge the assistance of the education authorities, colleges, teachers and students who contributed to the SCHOLAR programme and who evaluated these materials.

Grateful acknowledgement is made for permission to use the following material in the SCHOLAR programme:

The Scottish Qualifications Authority for permission to use Past Papers assessments.

The Scottish Government for financial support.

The content of this Study Guide is aligned to the Scottish Qualifications Authority (SQA) curriculum.

All brand names, product names, logos and related devices are used for identification purposes only and are trademarks, registered trademarks or service marks of their respective holders.

Contents

1 DNA and the Genome — 1

1. Structure and organisation of DNA — 3
2. Replication of DNA — 15
3. Gene expression — 27
4. Differentiation in multicellular organisms — 49
5. Structure of the genome — 67
6. Mutations — 73
7. Evolution — 93
8. Genomics — 109
9. End of unit test — 121

2 Metabolism and Survival — 131

1. Metabolic pathways — 133
2. Cellular respiration — 153
3. Metabolic rate — 167
4. Metabolism in conformers and regulators — 173
5. Maintaining metabolism — 189
6. Environmental control of metabolism — 199
7. Genetic control of metabolism — 219
8. End of unit test — 231

3 Sustainability and Interdependence — 241

1. Food supply — 245
2. Plant growth and productivity — 261
3. Plant and animal breeding — 283
4. Crop protection — 297
5. Animal welfare — 313
6. Symbiosis — 321
7. Social behaviour — 333

8 Components of biodiversity 353
9 Threats to biodiversity 361
10 End of unit test 375

Glossary   **384**

Answers to questions and activities   **395**

Unit 1: DNA and the Genome

1	**Structure and organisation of DNA**	**3**
	1.1 The structure of DNA	5
	1.2 The organisation of DNA in prokaryotes and eukaryotes	9
	1.3 Learning points	11
	1.4 Extension materials	12
	1.5 End of topic test	13
2	**Replication of DNA**	**15**
	2.1 DNA replication	17
	2.2 The polymerase chain reaction (PCR)	19
	2.3 Learning points	22
	2.4 Extended response question	23
	2.5 Extension materials	23
	2.6 End of topic test	25
3	**Gene expression**	**27**
	3.1 Introduction	29
	3.2 The structure and functions of RNA	29
	3.3 Transcription	32
	3.4 Translation	36
	3.5 One gene, many proteins	41
	3.6 Protein structure and function	42
	3.7 Learning points	43
	3.8 Extended response question	45
	3.9 End of topic test	45
4	**Differentiation in multicellular organisms**	**49**
	4.1 Introduction	51
	4.2 Meristems	51
	4.3 Stem cells	56
	4.4 Embryonic stem cells	57
	4.5 Tissue (adult) stem cells	58
	4.6 Differentiation	59
	4.7 Therapeutic use of stem cells	59

UNIT 1. DNA AND THE GENOME

    4.8 Research involving stem cells 60
    4.9 Ethical issues regarding stem cells 60
    4.10 Learning points 62
    4.11 Extended response question 63
    4.12 Extension materials 63
    4.13 End of topic test 65

5 Structure of the genome **67**
    5.1 Introduction to the Genome 68
    5.2 The genome 68
    5.3 Learning points 70
    5.4 Extension materials 71
    5.5 End of topic test 72

6 Mutations **73**
    6.1 Single gene mutations 75
    6.2 Chromosome structure mutations 86
    6.3 The importance of mutations and gene duplication 89
    6.4 Learning points 89
    6.5 Extended response question 90
    6.6 Extension materials 90
    6.7 End of topic test 91

7 Evolution **93**
    7.1 Evolution 95
    7.2 Gene transfer 96
    7.3 Selection 96
    7.4 Speciation 101
    7.5 Learning points 105
    7.6 End of topic test 106

8 Genomics **109**
    8.1 Genomic sequencing 111
    8.2 Phylogenetics 112
    8.3 Comparative genomics 114
    8.4 Personal genomics 115
    8.5 Learning points 116
    8.6 Extension materials 117
    8.7 End of topic test 120

9 End of unit test **121**

Unit 1 Topic 1

Structure and organisation of DNA

Contents

1.1 The structure of DNA 5
1.2 The organisation of DNA in prokaryotes and eukaryotes 9
1.3 Learning points 11
1.4 Extension materials 12
1.5 End of topic test 13

Prerequisites

You should already know that:

- DNA takes the form of a double-stranded helix;
- the two strands of DNA are held together by complementary base pairs;
- DNA contains the four bases A, T, G and C which make up the genetic code.

UNIT 1. DNA AND THE GENOME

Learning objective

By the end of this topic, you should be able to:

- appreciate that DNA is found in all living organisms and is the chemical which carries hereditary information;
- understand that the sequence of bases within DNA is the genetic code;
- appreciate that the sum total of all the DNA bases is the genome;
- describe the structure of a nucleotide;
- describe the structure of DNA;
- describe how nucleotides combine to form a backbone and that base pairing holds the two strands together, forming a double helix;
- describe the base pairing rule;
- name the bonds which hold bases together;
- explain that the two strands of DNA lie anti-parallel to each other and are read in different directions;
- understand that DNA can be organised in a number of ways, either circular or linear;
- state that circular forms of DNA can be found in prokaryotes, mitochondria and chloroplasts;
- state that bacteria and yeast also contain smaller rings of DNA, called plasmids;
- state that the linear form of DNA found in eukaryotes is tightly packaged with associated proteins called histones.

1.1 The structure of DNA

In 1953, the structure of the DNA molecule was explained for the first time by two scientists, James Watson and Francis Crick, at the Cavendish Laboratory in Cambridge.

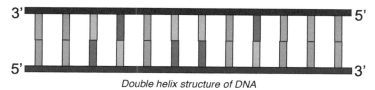

Double helix structure of DNA

They found that a molecule of DNA consists of two strands of repeating units called nucleotides. Nucleotides are composed of phosphate, deoxyribose sugar and a base.

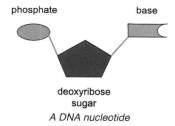

A DNA nucleotide

There are four bases in DNA: (A) adenine, (T) thymine, (G) guanine and (C) cytosine. Each nucleotide has a different base.

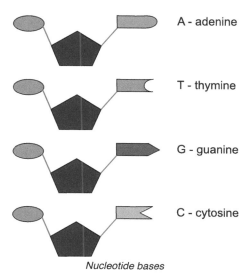

A - adenine

T - thymine

G - guanine

C - cytosine

Nucleotide bases

© HERIOT-WATT UNIVERSITY

UNIT 1. DNA AND THE GENOME

These bases follow the base pairing rules: A always pairs with T, and G always pairs with C. The bases are held together by hydrogen bonds.

DNA nucleotide		DNA nucleotide
A	pairs with	T
T	pairs with	A
G	pairs with	C
C	pairs with	G

Base pairing rules

The phosphate of one nucleotide is attached by a strong chemical bond to the deoxyribose sugar of the next. This forms the sugar-phosphate backbone of the DNA.

DNA takes the form of a **double helix**. DNA is made up of two **antiparallel** strands. This means the strands run in opposite directions to each other. The carbon atoms on the deoxyribose sugar are numbered as shown below.

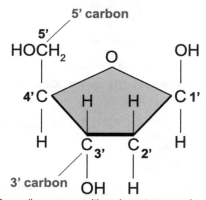

Deoxyribose sugar with carbon atoms numbered

Using this system, each end of the DNA can be labelled to show the antiparallel strands. The structure of DNA is shown as follows.

TOPIC 1. STRUCTURE AND ORGANISATION OF DNA

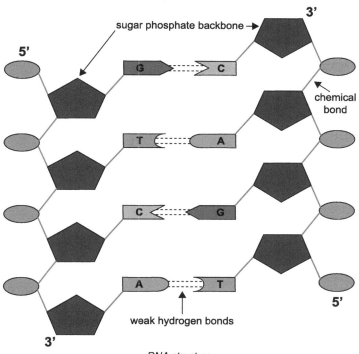

DNA structure

The structure of DNA: Interactive 3D model Go online

An activity that shows the double helix structure of DNA as an interactive 3D model is available in the online materials at this point. The following illustration gives an idea of what to expect.

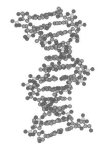

Double helix structure of DNA - molecular model

The structure of DNA: Questions

Go online

Q1: Complete the diagram using the words from the list.

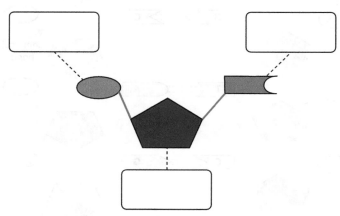

Word list: base, deoxyribose sugar, phosphate.

....

Q2: Complete the blanks using the base pairing rules.

- The nucleotide guanine pairs with _____.
- The nucleotide thymine pairs with _____.
- The nucleotide cytosine pairs with _____.
- The nucleotide adenine pairs with _____.

Q3: What shape is the DNA molecule?

....

Q4: What type of bonding holds two DNA strands together?

....

Q5: Which three components make up a DNA nucleotide?

1.2 The organisation of DNA in prokaryotes and eukaryotes

Prokaryotes are organisms which lack a true membrane-bound nucleus. Bacteria are an example of a prokaryote. Their DNA is found in the cytoplasm of the cell.

Eukaryotes are organisms which have a membrane bound nucleus that stores their genetic material. Animals, plants and fungi are examples of eukaryotes.

DNA is a double-stranded molecule that can either be circular or linear.

Circular and linear DNA

Prokaryotes have a large circular chromosome. They may also have smaller rings of DNA called **plasmids**.

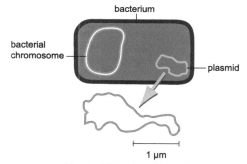

A bacterial (prokaryotic) cell

Circular plasmids may also be found in yeast (a fungus), which is classified as a eukaryote.

In eukaryotes, DNA is found tightly coiled into linear chromosomes. DNA is also found within mitochondria (mtDNA) where it forms circular chromosomes. It is sometimes described as the smallest chromosome and is inherited from the mother in humans.

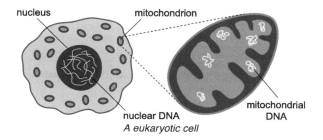

A eukaryotic cell

DNA can also be found within the chloroplasts of plant cells. It is usually larger than mitochondrial DNA and takes the form of circular chromosomes containing the genes involved in the photosynthetic process. Where circular DNA is found in eukaryotes, it is thought that it has been incorporated from early bacteria or prokaryotes.

Typically, the DNA content of a single human cell, if completely unravelled, would measure around two metres in length. This DNA must be packaged so it can fit inside the nucleus. DNA found in the linear chromosomes of the nucleus of eukaryotes is tightly coiled and packaged with associated proteins called histones. This is shown in the diagram below.

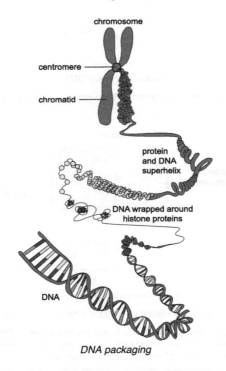

DNA packaging

The organisation of DNA in prokaryotes and eukaryotes: Questions   Go online

Q6: In which type of organism is circular DNA mainly found?
....

Q7: In which structures of eukaryotic cells might circular DNA be found?
....

Q8: What is the term for sections of extra-chromosomal DNA sometimes found in organisms?

1.3 Learning points

Summary

- DNA encodes hereditary information in a chemical language.
- All cells store their genetic information in the base sequence of DNA.
- The structure of a DNA nucleotide is composed of deoxyribose sugar, a phosphate and a base.
- Nucleotides bond to form a sugar-phosphate backbone.
- Base pairs (adenine, thymine, guanine and cytosine) hold the two strands together by hydrogen bonds, forming a double helix.
- Adenine always pairs with thymine, and guanine always pairs with cytosine.
- DNA is a double-stranded, antiparallel (one strand goes from 3' to 5', the other from 5' to 3') structure with a deoxyribose and phosphate backbone held together by internal base pairs.
- The DNA molecule can be circular or linear.
- Circular chromosomal DNA and plasmids are found in prokaryotes.
- Circular plasmids are found in yeast.
- Circular chromosomes are in the mitochondria and chloroplasts of eukaryotes.
- The DNA found in the linear chromosomes of the nucleus of eukaryotes is tightly coiled and packaged with associated proteins called histones.
- The sum total of all genetic material is the genome.
- Chromosomes in eukaryotes are contained within a membrane-bound nucleus.

1.4 Extension materials

The material in this section is not examinable. It includes information which will widen your appreciation of this section of work.

Extension materials: The Discovery of DNA

The significance of DNA and its role in hereditary can be traced from the work of Griffiths, who in 1928 demonstrated the "transforming principle" in bacteria. He, and later others (Avery, McCartney & McLeod, 1944), would show this "transforming principle" to be DNA.

Later, in the 1950s, Hershey & Chase, working with bacteriophage and radioactive forms of phosphorus and sulfur, would confirm DNA as the genetic material and eliminated protein as the carrier of genetic information.

Edwin Chargaff, using paper chromatography and ultraviolet spectroscopy techniques, demonstrated two findings, now known as Chargaff's rules: firstly, that adenine and thymine always occur together, and similarly that cytosine and guanine pair up - this is called base pairing; secondly, that DNA sequences vary between species.

In the early 1950s, work by Maurice Wilkins and Rosalind Franklin uncovered some characteristic features of the DNA molecule. Using a method called X-ray crystallography, it was shown that the molecule had a helical structure. Using this, and other evidence, Francis Crick and James Watson were able to construct the model of DNA that we recognise today.

The main features of the model are that it not only shows the configuration of the molecule, but it allows for the explanation of two processes: the first is the mechanism for DNA replication (semi-conservative) and the second is how it codes for proteins.

The story of the discovery of the structure of DNA, and an excellent book about how science works, is: *The Double Helix* by James D. Watson, published by Penguin Books.

1.5 End of topic test

End of Topic 1 test Go online

Q9: What is the term for the sum total of DNA in an organism?

....

Q10: What is the complementary base sequence for the following DNA sequence: ATGGACTTTAGGT?

....

Q11: A DNA molecule contains 1500 bases, of which 450 are adenine. What percentage of the bases in the DNA molecule are guanine?

....

Q12: The following diagram shows part of a DNA molecule.

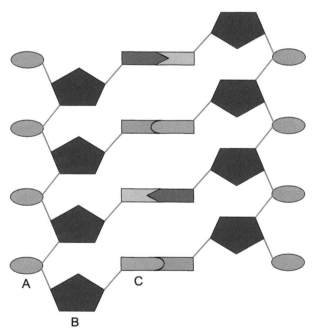

Which of the following correctly identifies **base**, **phosphate** and **sugar**, in that order?

1. A, B, C
2. A, C, B
3. B, A, C
4. C, A, B

UNIT 1. DNA AND THE GENOME

Q13: What is the name of the sugar that is found in DNA?

Q14: Base pairs in a DNA molecule are linked by weak bonds. What is the name for these?

Q15: What name is given to the backbone of DNA?

Q16: The diagram below illustrates the structure of a DNA molecule.

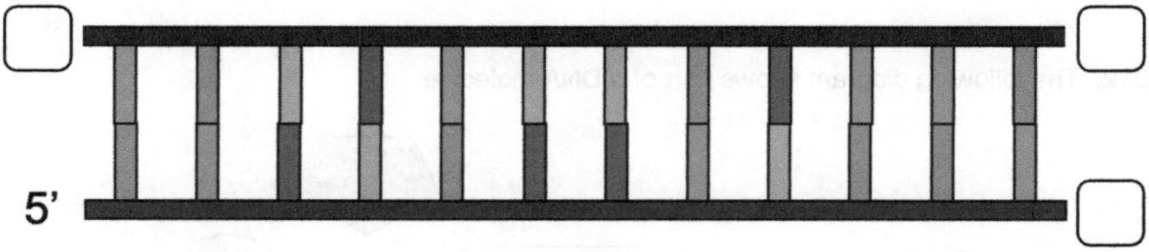

5'

Fill in the missing labels to complete the diagram.

Q17: Name two sources of circular DNA in eukaryotes.

Q18: Name the substance which DNA is packaged with in the nucleus of eukaryotes.

Unit 1 Topic 2

Replication of DNA

Contents

2.1 DNA replication 17
2.2 The polymerase chain reaction (PCR) 19
2.3 Learning points 22
2.4 Extended response question 23
2.5 Extension materials 23
2.6 End of topic test 25

Prerequisites

You should already know that:

- chromosomes (and therefore DNA) are replicated during mitosis;
- the two strands of DNA are held together by complementary base pairs;
- DNA contains the four bases A, T, G and C which make up the genetic code.

UNIT 1. DNA AND THE GENOME

Learning objective

By the end of this topic, you should be able to:

- explain the function of DNA polymerase, when it acts, and what conditions are necessary for it to function;
- describe how DNA replicates in terms of DNA unwinding and DNA polymerase adding complementary nucleotides;
- explain the importance and significance of adding new nucleotides to the 3' end of an existing DNA chain;
- state that replication occurs at various points on a DNA molecule;
- outline the process where DNA polymerase can act continuously on the leading strand, but discontinuously on the lagging strand;
- describe the action and significance of the enzyme ligase;
- state the purpose of the polymerase chain reaction (PCR);
- describe the action and purpose of primers in PCR;
- describe the sequence of PCR in terms of heating DNA, adding primers, and cooling DNA;
- explain why heat tolerant DNA polymerase is required;
- explain the significance and outcome of 'cycling' sequences;
- describe the practical applications of PCR.

TOPIC 2. REPLICATION OF DNA

2.1 DNA replication

DNA replication takes place prior to cell division. The replication of DNA is semi-conservative. Each strand acts as a template for the synthesis of a new DNA molecule by the addition of complementary base pairs, thereby generating a new DNA strand that is the complementary sequence to the parental DNA. Each daughter DNA molecule ends up with one of the original strands and one newly synthesised strand.

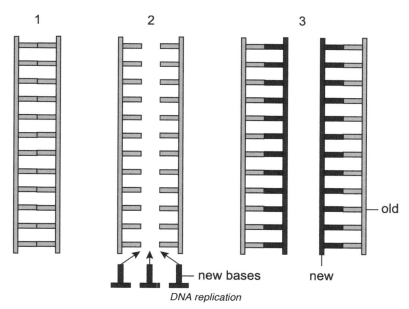

DNA replication

DNA replication is an enzyme controlled process which relies on the activities of **DNA polymerase** and **DNA ligase**.

- **DNA polymerase**: adds DNA nucleotides, using complementary base pairing, to the deoxyribose (3') end of the new DNA strand which is forming.
- **DNA ligase**: joins fragments of DNA together.

Before replication begins, there must be a pool of free nucleotides present; however, DNA polymerase cannot start adding nucleotides on its own. Short sections of RNA nucleotides called **primers** are added to the DNA and the enzyme extends from them. A primer is a short strand of nucleotides which binds to the 3' end of the template DNA strand allowing polymerase to add DNA nucleotides.

Due to the action of the enzyme DNA polymerase, the two strands of DNA are copied differently. The **leading strand** is made continuously while the **lagging strand** is made in fragments, which are then joined together.

DNA replication: Steps

The following provides a summary of the steps involved in DNA replication.

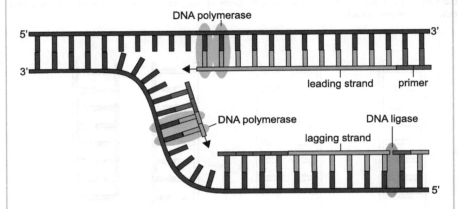

- DNA is unwound and hydrogen bonds between bases are broken to form two template strands.
- two replication forks form and open the double-strand in opposite directions, exposing the bases.
- on the leading strand, a primer binds to the DNA and DNA polymerase adds nucleotides to the 3' end. DNA polymerase catalyses the formation of a chemical bond between nucleotides and continues to add nucleotides to the 3' end of the growing strand.
- on the lagging strand, a primer binds to the DNA once it is exposed and DNA polymerase adds nucleotides to the 3' end. As more DNA is exposed, a new primer is added. DNA polymerase extends the new strand from this primer until it meets the previous fragment. The old primer is replaced by DNA and the enzyme DNA ligase joins the fragments together. As the DNA unzips further, another fragment will be made and connected to the previous one.

2.2 The polymerase chain reaction (PCR)

PCR can be used to amplify a desired DNA sequence of any origin (virus, bacteria, plant or animal) millions of times in a matter of hours. It is especially useful because:

- it is highly specific;
- it is easily automated;
- it is capable of amplifying minute amounts of sample.

To amplify a target DNA sequence, several components are required:

- buffer;
- nucleotides;
- **primers**;
- Taq polymerase;
- template DNA.

The section of DNA which is to be amplified must be added to the reaction mixture. It acts as a template to copy from. The buffer keeps the reaction mixture at the correct pH to ensure the reaction will proceed.

Polymerase enzyme is found in all animals. It has an optimum temperature of 37°C. PCR requires polymerase to operate at high temperatures. *Thermus aquaticus* is a bacterium that lives in hot springs and hydrothermal vents. Taq polymerase, an enzyme which adds nucleotides to DNA, was first isolated from this bacteria. It is special type of polymerase which is stable at high temperatures, having an optimum temperature of 70°C.

Polymerase can only add nucleotides to an existing strand of DNA, therefore, PCR requires primers. In PCR, primers are short strands of nucleotides which are complementary to specific target sequences at the two ends of the region of DNA to be amplified.

The PCR process involves repeated cycles of the following steps:

- the DNA molecule which is to be amplified is denatured by heating to between 92 and 98°C, breaking the hydrogen bonds between base pairs to separate the strands;
- the solution is cooled to between 50 and 65°C to allow the primers to bind to target sequences;
- the solution is heated to between 70 and 80°C for heat-tolerant DNA polymerase to replicate the region of DNA.

This cycle is usually repeated at least 30 times.

Polymerase chain reaction allows DNA to be amplified *in vitro*; this means it is performed outwith a living organism. The opposite, *in vivo*, means carried out within an organism.

The polymerase chain reaction (PCR): Stages

Go online

The following provides a summary of the stages of PCR.

Stage 1, Heating (92-98°C): DNA is heated to separate the strands.

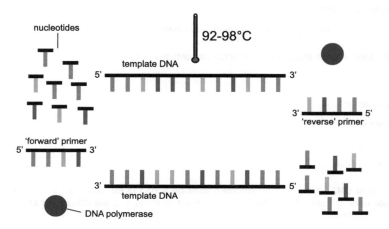

Stage 2, Annealing (50-65°C): DNA is then cooled to allow primers to bind to target sequences.

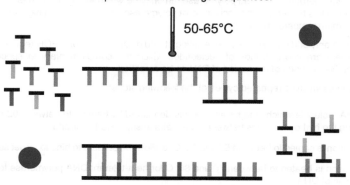

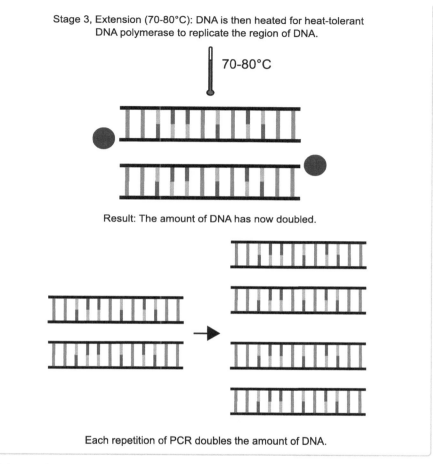

Stage 3, Extension (70-80°C): DNA is then heated for heat-tolerant DNA polymerase to replicate the region of DNA.

70-80°C

Result: The amount of DNA has now doubled.

Each repetition of PCR doubles the amount of DNA.

PCR has revolutionised many areas of research science. PCR is involved in the process of DNA sequencing and has allowed amplification of DNA from ancient sources, such as Neanderthal bones, enabling in-depth DNA analysis. PCR also has forensic applications, allowing minute quantities of DNA from a crime scene to be amplified, sequenced, and compared to DNA sequences from suspects. PCR has medical applications, for example the diagnosis of human immunodeficiency virus (HIV). Finally, PCR can be used to settle paternity disputes by amplifying and comparing a child's DNA to their potential father.

2.3 Learning points

Summary

- Prior to cell division, DNA polymerase replicates a DNA strand precisely using DNA nucleotides.
- DNA polymerase needs a primer to start replication.
- A primer is a short strand of nucleotides which binds to the 3' end of the template DNA strand allowing polymerase to add DNA nucleotides.
- DNA unwinds to form two template strands.
- DNA polymerase adds DNA nucleotides, using complementary base pairing, to the deoxyribose (3') end of the new DNA strand which is forming.
- This process occurs at several locations on a DNA molecule.
- DNA polymerase can only add DNA nucleotides in one direction resulting in the leading strand being replicated continuously and the lagging strand replicated in fragments.
- Fragments of DNA are joined together by ligase.
- The polymerase chain reaction (PCR) is a technique for the amplification of DNA *in vitro*.
- In PCR, primers are short strands of nucleotides which are complementary to specific target sequences at the two ends of the region of DNA to be amplified.
- PCR is a three step process:
    - heating (92-98°C) separates the DNA strands;
    - annealing (50-65°C) is the binding of the primers to target sequences;
    - extension (70-80°C) of the primers to complete the complementary strand is carried out by heat stable DNA polymerase.
- Repeated cycles of heating and cooling amplify this region of DNA.
- PCR can amplify DNA to help solve crimes, settle paternity suits and diagnose genetic disorders.

2.4 Extended response question

The activity which follows presents an extended response question similar to the style that you will encounter in the examination.

You should have a good understanding of DNA structure and replication before attempting the question.

You should give your completed answer to your teacher or tutor for marking, or try to mark it yourself using the suggested marking scheme.

Extended response question: DNA structure and replication

Give an account of DNA structure and replication. *(8 marks)*

2.5 Extension materials

The material in this section is not examinable. It includes information which will widen your appreciation of this section of work.

Extension materials: Meselson and Stahl's experiment

The mechanism of DNA replication is said to be semi-conservative. That is, after replication, each of the two resulting DNA molecules is composed of one original (or conserved) strand and one new strand. This hypothesis, put forward by Watson and Crick, was proved experimentally by Meselson and Stahl in the late 1950s. Their experiment involved growing a culture of the bacterium *Escherichia coli* in a growth medium containing heavy nitrogen (^{15}N). As the bacteria grew, they incorporated the heavy nitrogen into their nitrogenous bases. The bacteria were then inoculated into growth media containing light nitrogen (^{14}N) and three classes of DNA were subsequently extracted after the change to the light nitrogen medium.

The three classes of DNA were:

- parental DNA;
- first generation DNA;
- second generation DNA.

The results of the experiment showed that parental DNA grown in heavy medium was 'heavier' than when grown in light medium. First generation growth showed that the DNA was all of medium density. Lastly, the second generation showed DNA of both medium and light intensities. Second generation growth supported the semi-conservative model of DNA replication since there were two bands of growth (one with both conserved and new DNA, and a band of light DNA).

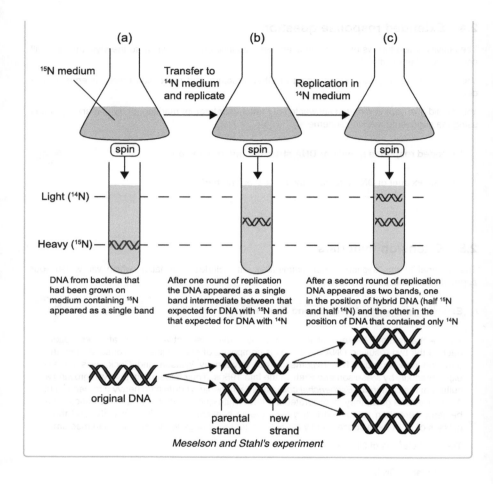
Meselson and Stahl's experiment

2.6 End of topic test

End of Topic 2 test Go online

Q1: Name the enzyme that is required to add nucleotides to a growing strand of DNA.

....

Q2: What name is given to the short strand of nucleotides added to DNA at the start of the replication process?

....

Q3: Name the enzyme that is required to join fragments of DNA together.

....

Q4: The following four stages occur during DNA replication:

a) Base pairing occurs between free nucleotides and each of the DNA strands.
b) The hydrogen bonds between DNA strands break.
c) The DNA molecules coil up to form double helices.
d) Nucleotides are bonded together by DNA polymerase.

Which of the following gives these stages in the correct order?

1. a, b, d, c
2. b, a, d, c
3. c, b, a, d
4. d, a, b, c

....

Q5: The DNA polymerase used in the polymerase chain reaction possesses a particular characteristic that makes it ideally suited to the purpose. What is it?

a) The enzyme is relatively stable at high temperatures.
b) The enzyme synthesises DNA at a very rapid rate.
c) The enzyme is very accurate at copying DNA from a template.
d) The enzyme can seal together fragments of DNA.

....

Q6: The steps involved in the polymerase chain reaction are given below.

a) Temperature of the reaction adjusted to 70-80°C.
b) Temperature of the reaction adjusted to 50-65°C.
c) The DNA strands separate.
d) Temperature of the reaction adjusted to 92-98°C.
e) Synthesis of DNA by the enzyme DNA polymerase.
f) Annealing of the primers to the single-stranded DNA.

Which of the following describes the correct order in which the steps would occur?

1. d, c, b, e, a, f
2. d, c, a, f, b, e
3. d, c, b, f, a, e
4. d, c, a, e, b, f

....

Q7: _____ occurs at 70-80°C.
Word list: extension, annealing, heating, exponential.

....

Q8: PCR leads to an _____ amplification of desired DNA sequences.
Word list: extension, annealing, heating, exponential.

....

Q9: Starting with a single molecule of DNA, the polymerase chain reaction was allowed to go through three complete cycles. How many molecules of DNA would be produced?

a) 4
b) 8
c) 16
d) 32

Unit 1 Topic 3

Gene expression

Contents

- 3.1 Introduction 29
- 3.2 The structure and functions of RNA 29
- 3.3 Transcription 32
- 3.4 Translation 36
- 3.5 One gene, many proteins 41
- 3.6 Protein structure and function 42
- 3.7 Learning points 43
- 3.8 Extended response question 45
- 3.9 End of topic test 45

Prerequisites

You should already know that:

- DNA carries the genetic information for making proteins;
- the base sequence of DNA determines the amino acid sequence in proteins;
- messenger RNA (mRNA) is a molecule which carries a copy of the code from the DNA in the nucleus to a ribosome in the cytoplasm;
- at the ribosome, proteins are assembled from amino acids.

UNIT 1. DNA AND THE GENOME

Learning objective

By the end of this topic, you should be able to:

- explain that the genetic code is universal to all forms of life;
- explain that the phenotype is a combination of the genotype and environmental factors;
- describe the structure of RNA;
- describe the function of mRNA, tRNA and rRNA;
- describe the process of transcription including the role of RNA polymerase and complementary base pairing;
- describe the process of RNA splicing;
- describe the process of translation, including the role of tRNA and ribosomes;
- state that codons are found on mRNA and anticodons are found on tRNA;
- name the bonds which hold amino acids together in a protein;
- explain how different proteins can be expressed from one gene;
- state that polypeptide chains fold to give the final structure of the protein;
- describe the role of hydrogen bonds and the interactions between amino acids in the 3D shape of a protein.

TOPIC 3. GENE EXPRESSION

3.1 Introduction

The many thousands of proteins that our cells use are synthesised inside the cells during a complex process involving DNA and RNA (the nucleic acids), as well as **ribosomes**. You will remember from previous studies that a protein is composed of amino acids joined together in a specific sequence.

The information to determine the sequence of amino acids in a protein is contained in the DNA in the nucleus of our cells. In this topic, we will study how the instructions for making a protein are transferred into the cytoplasm using RNA, and how a protein is actually constructed on the ribosomes.

Gene expression involves two major stages.

- The first process is **transcription**, during which the DNA is used to produce an RNA molecule that is called a primary transcript. This RNA has the same sequence as the gene. Human genes can be divided into **exons** and **introns**, but it is only the exons that carry the information needed for protein synthesis.

- The next stage of gene expression is known as **translation**, which allows amino acids to come together in a certain order at the **ribosome**, where they form a polypeptide chain.

DNA	makes	mRNA	makes	protein
Process:	transcription		translation	
Occurs in:	the nucleus		cytoplasm	

Protein synthesis summary

3.2 The structure and functions of RNA

Ribonucleic acid (RNA) provides a bridge between DNA and protein synthesis.

RNA consists of nucleotides that are composed of phosphate, ribose sugar and a base.

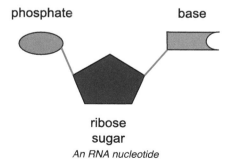

An RNA nucleotide

Important points to remember about RNA structure are that:

- RNA nucleotides contain the sugar ribose;
- RNA has the base (U) uracil rather than (T) thymine (as found in DNA);
- RNA molecules are usually single-stranded.

There are three main types of RNA involved in protein synthesis:

1. **mRNA** (messenger RNA), which carries a copy of the DNA code from the nucleus to the ribosome;
2. **tRNA** (transfer RNA), which are molecules found in the cytoplasm that become attached to specific amino acids, bringing them to the ribosomes where they are joined together;
3. **rRNA** (ribosomal RNA), which forms a complex with protein molecules to make the ribosome.

Messenger RNA (mRNA)	For the synthesis of a protein, the particular sequence of bases on the DNA is first transcribed into the complementary sequence of mRNA. This messenger RNA can then carry the information for a protein through the nuclear envelope to the sites of protein synthesis (the ribosomes). Each triplet of bases on the mRNA molecule is called a codon and codes for a specific amino acid.
Transfer RNA (tRNA)	This type of RNA is responsible for the transport and transfer of individual amino acids during protein synthesis. Amino acids are transported by specific tRNA molecules, which recognise the genetic code presented by the mRNA. The three bases exposed at the bottom form the anticodon. This is the complementary base sequence to the base sequence on mRNA coding for a particular amino acid.
Ribosomal RNA (rRNA)	This type of RNA is bound to structural proteins to form a ribosome. The ribosome is used in the synthesis of proteins.

The three forms of RNA

TOPIC 3. GENE EXPRESSION

The structure and functions of RNA: Questions

Go online

Q1: Complete the diagram using the labels from the list.

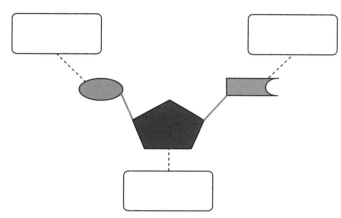

Labels: base, phosphate, ribose sugar.

....

Q2: Complete the table using the properties of RNA and DNA from the list.

	RNA	DNA
Structure		
Preferred form		
Number of types		
Present in		
Bases		

Properties: >1; 1; adenine, cytosine, guanine and thymine; adenine, cytosine, guanine and uracil; double-stranded; double helix; not a double helix; single-stranded; the cytoplasm and the nucleus; the nucleus.

Q3: What are the components of an RNA nucleotide?

....

Q4: List the four types of bases found in an RNA molecule.

....

Q5: Name the three types of RNA found in the cell.

....

32 UNIT 1. DNA AND THE GENOME

Q6: What are main functional differences between mRNA and tRNA?

....

Q7: Nucleotides are the building blocks of:

a) DNA only
b) RNA only
c) both DNA and RNA
d) neither DNA nor RNA

3.3 Transcription

Transcription is the first step in protein synthesis. Information from DNA is copied into an RNA molecule, a process which takes place in the nucleus. The RNA polymerase enzyme moves along the DNA, unwinding the double helix and breaking the hydrogen bonds that hold the base pairs together. Free RNA nucleotides bond with the complementary base pairs on the DNA. The base pairing rules are summarised in the table below. The RNA nucleotides are held in place by hydrogen bonds while strong bonds form between the phosphate of one nucleotide and the ribose sugar of the adjacent nucleotide. When transcription is complete, the RNA polymerase enzyme and the mRNA strand that has been constructed are released. The mRNA that has been produced at this stage is known as the primary transcript.

DNA nucleotide		RNA nucleotide
A	pairs with	U
T	pairs with	A
G	pairs with	C
C	pairs with	G

RNA base pairing rules

© HERIOT-WATT UNIVERSITY

TOPIC 3. GENE EXPRESSION

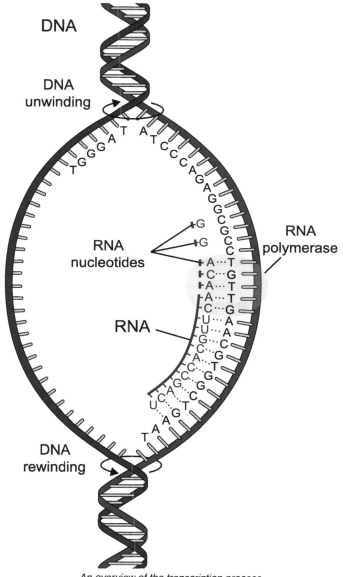

An overview of the transcription process

A section of DNA containing the genetic instructions for the production of a protein.

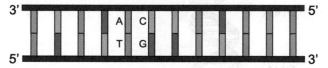

RNA polymerase unwinds the DNA and breaks the hydrogen bonds between the bases. This causes the strands to separate and expose their bases. Free RNA nucleotides find and align with complementary DNA nucleotides by hydrogen bonding. Strong chemical bonds form between the sugar of one RNA nucleotide and the phosphate of the next.

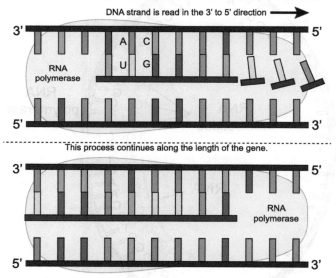

This process continues along the length of the gene.

The weak hydrogen bonds between the DNA and RNA bases break, allowing the mRNA to separate from the DNA and then move away from the DNA. The weak hydrogen bonds between the DNA strands re-unite and the molecule winds up into a double helix again.

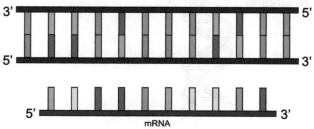

The steps involved in RNA transcription

TOPIC 3. GENE EXPRESSION

After a eukaryotic cell transcribes a protein coding gene, the RNA transcript, called the primary transcript, is processed. One type of processing is called **RNA splicing**. This takes place in the nucleus, after which the mature mRNA is released into the cytoplasm where ribosomes translate the mRNA transcript into protein.

RNA splicing forms a mature mRNA transcript. The **introns** of the primary transcript are non-coding regions and are removed. The **exons** are coding regions and are joined together to form the mature transcript. The order of the exons is unchanged during splicing.

RNA splicing: Steps

Go online

The following provides a summary of the steps involved in RNA splicing.

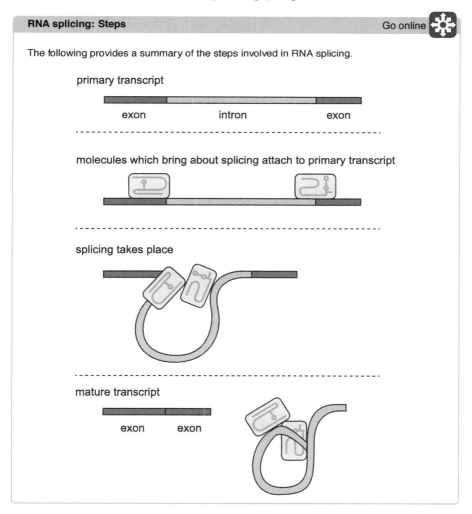

> **Transcription: Questions** — Go online
>
> **Q8:** Name the enzyme which is responsible for producing mRNA.
>
> ....
>
> **Q9:** Give the location of transcription.
>
> ....
>
> **Q10:** Name the bonds between DNA bases which must be broken during transcription.
>
> ....
>
> **Q11:** What name is given to the mRNA molecule which has just been released from the DNA?
>
> ....
>
> **Q12:** Name the process which removes introns from the primary transcript.

3.4 Translation

Once the DNA in a gene has been transcribed into mRNA, **translation** can take place. First, the mRNA molecules pass through the nuclear pores. Translation of mRNA into protein takes place on **ribosomes** in the cytoplasm and requires a second type of RNA, **transfer RNA (tRNA)**.

> **tRNA** — Go online
>
> The following provides structural information about a tRNA molecule, in this case with the amino acid serine attached to it.
>
>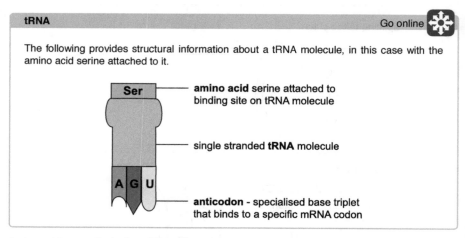

Amino acids are attached to tRNA at the amino acid attachment site at the top of the molecule. Unlike mRNA, there are some regions of base pairing in a tRNA molecule. The three bases exposed at the base of the tRNA molecule is called the **anticodon**. (Each group of three bases on the mRNA which codes for an amino acid is called a **codon**).

As an mRNA molecule passes through a ribosome, each **codon** is translated into an amino acid. The genetic code table indicates which amino acid corresponds to each mRNA codon. The tRNA molecule carrying the complementary anticodon binds briefly to the mRNA codon. The amino acid attached to the tRNA is then added to the polypeptide chain being synthesised. Amino acids are joined together by strong peptide bonds. After the amino acid has been added to a polypeptide chain during translation, the tRNA is free to pick up another amino acid in the cytoplasm.

Second letter

		U	C	A	G		
First	U	phe	ser	tyr	cys	U	**Third**
letter	U	phe	ser	tyr	cys	C	**letter**
	U	leu	ser	stop	stop	A	
	U	leu	ser	stop	trp	G	
	C	leu	pro	his	arg	U	
	C	leu	pro	his	arg	C	
	C	leu	pro	gln	arg	A	
	C	leu	pro	gln	arg	G	
	A	ile	thr	asn	ser	U	
	A	ile	thr	asn	ser	C	
	A	ile	thr	lys	arg	A	
	A	met	thr	lys	arg	G	
	G	val	ala	asp	gly	U	
	G	val	ala	asp	gly	C	
	G	val	ala	glu	gly	A	
	G	val	ala	glu	gly	G	

Genetic code table

When the polypeptide chain is completed, it is released from the ribosome. Further processing, such as folding and binding to other polypeptide chains, results in the formation of a mature protein. The mRNA molecule is usually reused to produce more identical polypeptide chains.

There are three codons that do not code for amino acids: UGA, UAA and UAG. These codons are known as stop codons and signal where translation ends. The genetic code also includes start codons where translation begins. In eukaryotes this is almost always AUG, which also codes for the amino acid methionine.

Translation: Steps

Go online

As a molecule of mRNA slides through a ribosome, **codons** are translated into amino acids, one by one

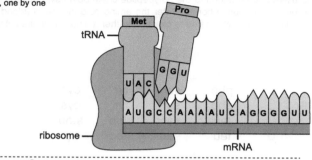

tRNA molecules have a specific **anticodon** at on end and a certain **amino acid** at the other. A tRNA adds its amino acid cargo to a growing polypeptide chain when the anticodon bonds to a complementary codon on the mRNA

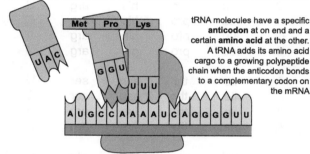

Strong **peptide bonds** form between amino acids in the growing peptide chain. Each tRNA molecule becomes attached to another molecule of its amino acid ready to repeat the process.

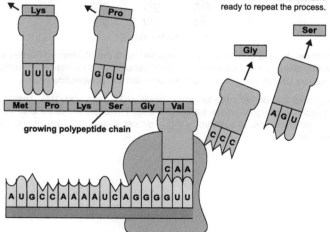

TOPIC 3. GENE EXPRESSION

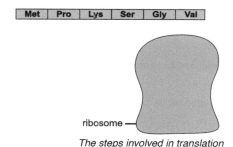

The mRNA can be reused to produce more identical polypeptides.

The steps involved in translation

Translation: Questions   Go online

The following steps describe the role of messenger RNA in the cell, providing a summary of protein synthesis transcription to translation.

Q13: The following steps describe the role of messenger RNA in the cell, providing a summary of protein synthesis from transcription to translation. List the steps in the correct order.

- Hydrogen bonds are formed between the first codon of the mRNA and the complementary anticodon on a tRNA
- Hydrogen bonds form between the bases on the RNA and the DNA nucleotides
- The second tRNA binds to the mRNA
- The first tRNA leaves the ribosome, and another tRNA enters and base-pairs with the mRNA
- The double-stranded DNA unwinds, hydrogen bonds in the DNA break and the DNA strands separate
- A second peptide bond is then formed. The process continues, with the ribosome moving along the mRNA
- The mRNA leaves the nucleus and enters the cytoplasm
- A peptide bond forms between the amino acids carried by the tRNA molecules
- A ribosome attaches to the mRNA. Two transfer RNA (tRNA) molecules are also contained within the ribosome
- As each mRNA codon is exposed, incoming tRNA pairs with it and polypeptide synthesis continues until completed
- When synthesis of the mRNA is completed, the mRNA separates from the DNA
- An RNA nucleotide binds to a complementary nucleotide on one of the DNA strands
- The RNA nucleotides are linked together to form messenger RNA (mRNA)

....

Q14: What name is given to the three bases exposed at the bottom of a tRNA molecule?
....
Q15: Where does translation take place?
....
Q16: Which cellular organelle is required for translation?
....
Q17: Name the bond which holds amino acids together in a protein.
....
Q18: Three mRNA codons do not code for amino acids. What is their role in translation?

Q19: Complete the sequence of bases encoded in the mRNA and then determine the sequence of amino acids in the protein. Use the genetic code table below to help.

Second letter

First letter		U	C	A	G		Third letter
	U	phe	ser	tyr	cys	U	
	U	phe	ser	tyr	cys	C	
	U	leu	ser	stop	stop	A	
	U	leu	ser	stop	trp	G	
	C	leu	pro	his	arg	U	
	C	leu	pro	his	arg	C	
	C	leu	pro	gln	arg	A	
	C	leu	pro	gln	arg	G	
	A	ile	thr	asn	ser	U	
	A	ile	thr	asn	ser	C	
	A	ile	thr	lys	arg	A	
	A	met	thr	lys	arg	G	
	G	val	ala	asp	gly	U	
	G	val	ala	asp	gly	C	
	G	val	ala	glu	gly	A	
	G	val	ala	glu	gly	G	

DNA	C	A	C	A	G	T	G	T	T	T	G	T	C	C	G
mRNA																
protein																

3.5 One gene, many proteins

Until quite recently, there was a theory that stated "one gene, one protein". This, however, has been superseded. With the completion of the Human Genome Project it is now accepted that the human genome contains between 20,000 and 25,000 genes, and yet it is also accepted that there are in excess of one million proteins in humans. Clearly there must be some mechanism that allows the genes to be expressed in a variety of ways.

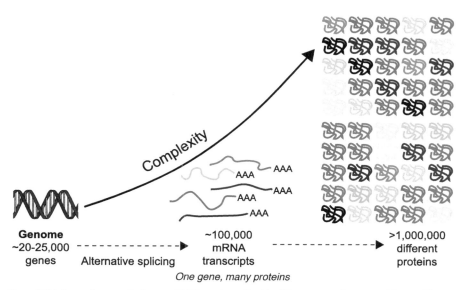

One gene, many proteins

The mRNA (sometimes called pre-mRNA) can be edited in different ways by assembling a different sequence of exons for translation. As a result, many different mature transcripts of mRNA can be derived from one section of DNA. This process is known as alternative splicing. As a result of alternative splicing, different mature mRNA transcripts are produced from the same primary transcript depending on which exons are retained.

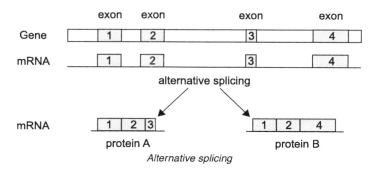

Alternative splicing

3.6 Protein structure and function

Proteins do not simply exist and function as strings of amino acids. After translation, the protein is folded to produce its final 3D shape. The folded polypeptide chains of a protein are held in place by hydrogen bonds and other interactions between individual amino acids.

Description	Diagram
Chain of amino acids linked by strong peptide bonds	
Polypeptide structure determined by weak hydrogen bonds	
Strong bonds form between special groups of amino acids	
More than one polypeptide makes up the final structure	

Protein structure

Different types of proteins have a variety of different functions within a living organism, examples of which are given in the following table.

Type of protein	Example
Structural	collagen, elastin
Contractile	actin and myosin in muscle cells
Hormones	insulin
Receptors	insulin receptor in liver cells (forming part of the structure of the plasma membrane)
Transport proteins	transporter of glucose into the cell
Defence proteins	immunoglobulins
Enzymes	lipase, pepsin, maltase

Protein functions

Protein structure and function: Questions

Go online

Q20: Complete the table by selecting from the listed images.

Description	Diagram
Chain of amino acids linked by strong peptide bonds	
Polypeptide structure determined by weak hydrogen bonds	
Strong bonds form between special groups of amino acids	
More than one polypeptide makes up the final structure	

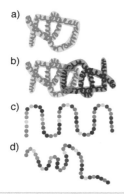

a)

b)

c)

d)

3.7 Learning points

Summary

- Gene expression involves the transcription and translation of DNA sequences.
- Only a fraction of the genes in a cell are expressed.
- RNA is single stranded and is composed of nucleotides containing ribose sugar, phosphate and one of four bases: cytosine, guanine, adenine and uracil.
- Transcription and translation involves three types of RNA (mRNA, tRNA and rRNA).
- Messenger RNA (mRNA) carries a copy of the DNA code from the nucleus to the ribosome.
- mRNA is transcribed from DNA in the nucleus and translated into proteins by ribosomes in the cytoplasm. Each triplet of bases on the mRNA molecule is called a codon and codes for a specific amino acid.

> **Summary continued**
>
> - Transfer RNA (tRNA) folds due to complementary base pairing. Each tRNA molecule carries its specific amino acid to the ribosome.
> - A tRNA molecule has an anticodon (an exposed triplet of bases) at one end and an attachment site for a specific amino acid at the other end.
> - Ribosomal RNA (rRNA) and proteins form the ribosome.
> - RNA polymerase moves along DNA unwinding the double helix and breaking the hydrogen bonds between the bases. RNA polymerase synthesises a primary transcript of mRNA from RNA nucleotides by complementary base pairing.
> - Uracil in RNA is complementary to adenine.
> - RNA splicing forms a mature mRNA transcript.
> - The introns of the primary transcript are non-coding regions and are removed.
> - The exons are coding regions and are joined together to form the mature transcript.
> - The order of the exons is unchanged during splicing.
> - tRNA is involved in the translation of mRNA into a polypeptide at a ribosome.
> - Translation begins at a start codon and ends at a stop codon.
> - Anticodons bond to codons by complementary base pairing, translating the genetic code into a sequence of amino acids. Peptide bonds join the amino acids together. Each tRNA then leaves the ribosome as the polypeptide is formed.
> - Different proteins can be expressed from one gene, as a result of alternative RNA splicing. Different mature mRNA transcripts are produced from the same primary transcript depending on which exons are retained.
> - Amino acids are linked by peptide bonds to form polypeptides.
> - Polypeptide chains fold to form the three-dimensional shape of a protein, held together by hydrogen bonds and other interactions between individual amino acids.
> - Proteins have a large variety of shapes which determines their functions.
> - Phenotype is determined by the proteins produced as the result of gene expression.
> - Environmental factors also influence phenotype.

TOPIC 3. GENE EXPRESSION

3.8 Extended response question

The activity which follows presents an extended response question similar to the style that you will encounter in the examination.

You should have a good understanding of protein synthesis before attempting the question.

You should give your completed answer to your teacher or tutor for marking, or try to mark it yourself using the suggested marking scheme.

Extended response question: Protein synthesis

Describe the role of RNA in protein synthesis. *(7 marks)*

3.9 End of topic test

End of Topic 3 test — Go online

Q21: The diagram shows two nucleotides that form part of a messenger RNA molecule. Which of the following correctly identifies phosphate, sugar and base, in order?

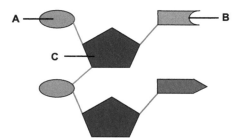

1. A, B and C
2. B, A and C
3. A, C and B
4. B, C and A

....

Q22: Which of the following correctly describes the structure of an RNA molecule?

a) It is single-stranded; it has deoxyribose in its backbone; it contains the base thymine.
b) It is double-stranded; it has ribose in its backbone; it contains the base uracil.
c) It is single-stranded; it has ribose in its backbone; it contains the base uracil.
d) It is double-stranded; it has deoxyribose in its backbone; it contains the base thymine.

Q23:

a) What is the messenger RNA (mRNA) sequence encoded by the following DNA sequence:
GCATTCATTGCA?

b) What name is given to the above process?

c) Name the enzyme required to carry out the above process.

Q24: Complete the following sentence by selecting the correct options from the word list.
The mRNA produced after transcription is called the _____; the _____ are removed, leaving only the _____ in the final _____.
Word list: exons, introns, mature transcript, primary transcript.

Q25: What name is given to the process which removes introns from mRNA?

Q26: What name is given to the process by which ribosomes use messenger RNA to produce a polypeptide chain?

Q27: The diagram below shows six transfer RNA (tRNA) molecules, each of which carries a different amino acid (indicated by the letters G, A, S, L, C and V). Part of a messenger RNA (mRNA) molecule is also shown.

tRNA molecules

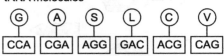

mRNA molecule

What is the sequence of amino acids encoded by the mRNA molecule?

TOPIC 3. GENE EXPRESSION

Q28: What name is given to the site indicated by letter X in the diagram below?

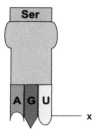

Q29: Name the process which allows different mature mRNA transcripts to be produced from the same primary transcript.

Q30: The following diagram shows part of a peptide chain from a protein.

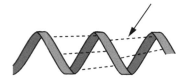

What type of bond is indicated by the arrow?

Unit 1 Topic 4

Differentiation in multicellular organisms

Contents

- 4.1 Introduction 51
- 4.2 Meristems 51
- 4.3 Stem cells 56
- 4.4 Embryonic stem cells 57
- 4.5 Tissue (adult) stem cells 58
- 4.6 Differentiation 59
- 4.7 Therapeutic use of stem cells 59
- 4.8 Research involving stem cells 60
- 4.9 Ethical issues regarding stem cells 60
- 4.10 Learning points 62
- 4.11 Extended response question 63
- 4.12 Extension materials 63
- 4.13 End of topic test 65

Prerequisites

You should already know that:

- stem cells have the potential to become different types of cell;
- stem cells are involved in growth and repair.

Learning objective

By the end of this topic, you should be able to:

- understand and explain the term differentiation;
- describe the role of plant meristems and how they generate new tissue during plant growth;
- describe the role of animal stem cells;
- describe some therapeutic uses of stem cells;
- describe the importance of stem cell research and take into account the ethical issues.

TOPIC 4. DIFFERENTIATION IN MULTICELLULAR ORGANISMS

4.1 Introduction

All living things are characterised by levels of organisation that are hierarchical. The cell is the lowest level of organisation that can exist independently. Multicellular organisms have cells organised into groups of cells called tissues, the next level of organisation. Tissues are formed from specialised cells that carry out a particular function. The columnar cells in the lining of the intestine, for example, are specialised for absorption of nutrients. Tissues can themselves become grouped together to form an organ. Most organs, such as the heart, lungs and liver, are also specialised for a certain function. The final level of organisation is the organ system where a group of organs work together at a particular function. Examples are the nervous system, the endocrine system and the vascular system.

Although the vast majority of cells contain identical genomes, cells within the same organism differ from one another because they express different genes and make different proteins.

Differentiation is the process by which cells or tissues undergo a change towards a more specialised function.

4.2 Meristems

Growth is restricted to specific regions (the meristems) of a plant, but it can occur throughout the plant's lifetime. In animals, growth can occur throughout the animal's body, but it stops when the animal reaches maturity. Animals do not have meristems; they are exclusive to plants.

Meristems are regions of unspecialised cells in plants that can divide (self-renew) and/or differentiate. These cells divide rapidly by **mitosis** to differentiate and form new plant tissues.

Apical meristems are located at the tips of the roots and shoots of a plant. The name is derived from the position at the tip, which is also known as the apex. They contain a cluster of actively dividing cells that increase the length of the plant. Therefore, in order for a plant to increase in length, it has to produce new cells at the apical meristems.

Differentiation: Phloem and xylem vessels

Go online

The following illustrates how cells differentiate to form phloem and xylem vessels.

1. Phloem sieve tubes transport carbohydrates throughout a plant.

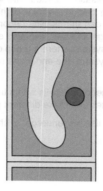

2. The cell beings to divide longitudinally and a second nucleus is formed.

3. After the cell division is complete, the nucleus in the sieve tube disintegrates and a companion cell is formed.

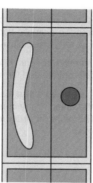

4. A cell wall forms between the sieve tube and companion cell. The end walls of the sieve tube become perforated and the cell contents die.

sieve tube
companion cell

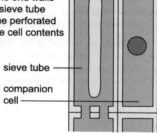

Differentiation within phloem vessels

1. Xylem vessels transport water throughout a plant and provide it with support.

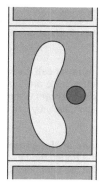

2. The nucleus disintegrates.

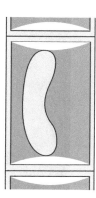

3. A strengthening material called **lignin** is deposited in a spiral on the inside of the cell wall. This gives support to the plant.

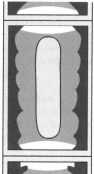

4. The cell contents die and the end walls disintegrate. This forms a hollow tube for water to be transported through.

lignin

xylem vessel

Differentiation within xylem vessels

Shoot growth: Zones

Go online

The following provides a summary of the zones involved in shoot growth.

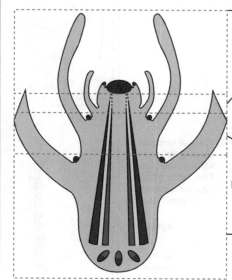

Zone of cell division and mitosis
New cells are formed in the shoot apical meristem when the undifferentiated cells multiply by cell division and mitosis

Zone of elongation and vacuolation
The undifferentiated cells produced in the meristem elongate and vacuoles form in them. This is the only zone in which growth occurs.

Zone of differentiation
No growth actually occurs in this zone, but undifferentiated cells differentiate. This forms specialised cells with particular functions.

Permanent tissues
The cells have fully differentiated to form specialised cells and do not undergo any more changes.

Root growth: Zones

Go online

The following provides a summary of the zones involved in root growth.

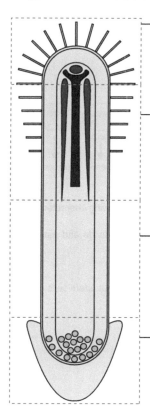

Permanent tissues
The cells have fully differentiated to form specialised cells and do not undergo any more changes

Zone of differentiation
No growth occurs in this zone but, the undifferentiated cells differentiate. This forms specialised cells with particular functions.

Zone of elongation and vacuolation
The undifferentiated cells produced in the meristem elongate and vacuoles form in them. This is the only zone in which the cells actually grow.

Zone of cell division and mitosis
New cells are formed in the root apical meristem when the undifferentiated cells multiply by cell division and mitosis. A lubricated root cap protects the apical meristem in the root tip. It also helps the root move through the soil.

4.3 Stem cells

Stem cells are unspecialised cells. When a stem cell divides, each new cell has the potential to remain a stem cell. This process is called **self-renewal**. In addition to self-renewal, stem cells can **differentiate** to become another type of cell with a more specialised function, such as a muscle cell, a red blood cell or a brain cell.

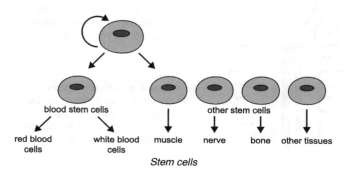

Stem cells

Stem cells are different from other body cells as they have the following characteristics:

1. Undifferentiated (unspecialised cell type), allowing them to divide and maintain a supply of stem cells for the body.
2. Found in all multicellular organisms.
3. Self-renewing and can differentiate; in some organs like the gut, stem cells regularly divide to repair and replace worn out or damaged tissues.

The two types of stem cells found in humans are:

1. Embryonic stem cells.
2. Tissue (adult) stem cells.

Stem cells: Questions   Go online

Q1: What do you understand by self-renewal in stem cells?

....

Q2: What is the unique property of stem cells which makes them different from a specialised cell?

....

Q3: Can you give examples of differentiated cells and their functions?

4.4 Embryonic stem cells

Most embryonic stem cells are derived from embryos that develop from eggs that have been fertilised, but before preimplantation.

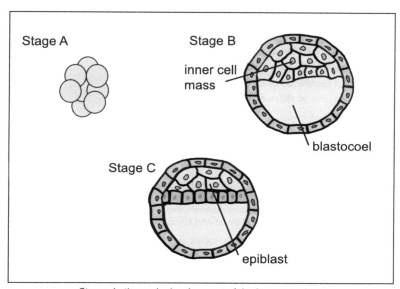

Stages in the early development of the human embryo

After the process of fertilisation, the zygote undergoes rapid cell division (stage A) and produces a multicellular ball called a blastula. The blastula contains a fluid-filled cavity called the blastocoel (stage B). In humans, the blastula becomes implanted in the uterus, and the cells of the inner cell mass begin to differentiate (stage C).

Embryonic stem cells are derived from embryos at the **blastocyst** stage. All the genes in embryonic stem cells can be switched on so these cells can differentiate into any type of cell. These cells are described as **pluripotent** as they can differentiate into all the cell types that make up the organism.

Human embryonic stem cells can be formed by transferring cells from a preimplantation-stage embryo into a plastic laboratory culture dish that contains a nutrient broth, known as culture medium. In more recent research into embryonic stem cells, scientists have reliably directed the differentiation of embryonic stem cells into specific cell types by using different culture conditions. They are able to use the resulting, differentiated cells to treat certain diseases.

> **Embryonic stem cells: Question** Go online
>
> **Q4:** Put the following stages of the process of using human embryonic stem cells (hESCs) to form specialised cells into the correct order.
>
> - Formation of mass of undifferentiated stem cells
> - Stem cell cultured in the laboratory
> - Embryo stem cell removed
> - Undifferentiated stem cells cultured in different culture conditions
> - Formation of specialised cells: nerve cell, muscle cell, gut cells
> - Early human embryo Blastocyst

4.5 Tissue (adult) stem cells

Early work on tissue stem cells started in the 1950s. Researchers discovered that bone marrow contains at least two kinds of stem cells. These were haematopoietic stem cells, which form all of the types of blood cells in the body, and bone marrow stem cells, which can generate bone, cartilage and fat cells that support the formation of blood and fibrous connective tissue.

Tissue stem cells are involved in the growth, repair and renewal of the cells found in that tissue. They are multipotent. Tissue stem cells are multipotent as they can differentiate into all of the types of cell found in a particular tissue type. For example, blood stem cells located in bone marrow can give rise to all types of blood cell.

Tissue stem cells have been identified in many organs and tissues, including brain, bone marrow, peripheral blood, blood vessels, skeletal muscle and skin. It is also worth knowing that these stem cells are also found in foetuses and babies. Although found in many types of tissues, only a very small number of stem cells actually occur in these tissues.

4.6 Differentiation

Cellular differentiation is the process by which a cell develops more specialised functions by expressing the genes characteristic for that type of cell. For example, a white blood cell only expresses genes which relate to its function, such as those which produce antibodies.

Cellular differentiation is the result of gene expression, which is under the influence of many factors. Not all cells complete the process of differentiation; some cells pause at the stage where they can still undergo duplication. This allows them to generate replacement cells which may die or be damaged - it allows for growth and repair. These cells are known as tissue (or adult) stem cells.

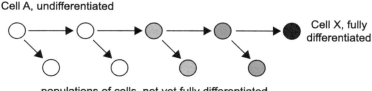

Differentiation

The original cell has the power to differentiate into several different varieties of cell. Cells differentiate gradually over several rounds of division until the final differentiated cells can no longer reproduce themselves.

4.7 Therapeutic use of stem cells

Some types of stem cells have been used in medicine for a number of years to repair damaged or diseased organs. Some examples are listed below:

- Skin: a rich source of tissue stem cells. Patients with serious burns can be treated using a technique which grows new skin in the lab from skin stem cells.

- Blood: a type of stem cell found in the bone marrow is capable of giving rise to all of the different types of blood cells. Bone marrow transplants have been carried out for many years as a treatment for diseases such as leukaemia and other blood disorders.

- Cornea: corneal stem cells are removed and cultured in a laboratory. They are then transplanted onto the diseased cornea to repair it.

4.8 Research involving stem cells

Almost all animals contain a very small population of cells that retain the ability to reproduce daughter cells which will, in turn, replace differentiated tissues that have become worn, diseased or damaged. However, the power of the stem cell to regenerate could be potentially dangerous and, if poorly regulated, give rise to many forms of cancer. An understanding of how stem cells regulate their own growth and development is therefore extremely important.

To begin studying stem cells, it is necessary to develop a stem cell line. A stem cell line is a group of constantly dividing cells from a single parent group of stem cells. Stem cell lines are grown in culture dishes, allowing them to divide and grow as undifferentiated cells for many years.

Stem cell research allows scientists to discover more information about key cell processes such as growth, differentiation and gene regulation. Stem cells can also be used to study how diseases develop. Stem cells may be a viable alternative to animals for testing new drugs in the future. The potential benefits of using stem cells in medicine seem endless. In the future, scientists hope that stem cells will be used to cure conditions such as Alzheimer's disease, Parkinson's disease, diabetes, traumatic spinal cord injury, vision and hearing loss, Duchenne's muscular dystrophy, stroke and heart disease.

4.9 Ethical issues regarding stem cells

Stem cell research and therapy are regulated in this country by the Human Fertilisation and Tissue Authority, and the Human Tissue Authority. Researchers and clinicians have to work within strict guidelines outlined in the Human Fertilisation and Embryology Act 1990, which was revised in 2000 and updated in 2008. In some other countries, all work in this area is outlawed.

Stem cell research and the sourcing of stem cells have produced a great deal of argument and discussion. Many, from principally religious and moral stand-points, have argued against this work. This is because they believe that life begins at the point of fertilisation and that the zygote or **blastocyst** should be thought of, and treated, as if it was the same as a living being. Currently, embryos up to 14 days may be used. Other groups, such as patients awaiting treatment, are supportive of stem cell research.

Other areas of research are attempting to grow stem cells from adult and differentiated tissue using a technique known as **induced pluripotentency**. In this process, adult differentiated cells are taken and re-engineered back to embryonic-like cells. This could be a way of expanding research in stem cells by avoiding the ethical issues associated with the use of embryonic stem cells.

A study of ethics

The ethical matrix (designed by Professor Ben Mepham, Centre for Applied Bioethics at the University of Nottingham) is a tool to help people analyse an ethical issue and make an informed choice. It is based on three key ethical principles (for further information on these see Mepham: *Bioethics: an introduction for the Biosciences* Oxford University Press (2008), now in 2nd edition):

1. Wellbeing: the safety, welfare and health of an individual or group.
2. Autonomy: an individual's right to be free to choose and make their own decisions.
3. Justice: to what extent a situation is just or fair for an individual or group.

You are asked to consider the following questions with reference to the information in the ethical matrix to help explore your own opinions and feelings using, as far as possible, the evidence here and any other source you feel appropriate.

a) What do you think might be the priority of each of the interest groups?

b) In what way do you think that the three principals apply to each interest group?

c) To what extent do you think others might agree or disagree with you?

d) Might your decision be influenced by the thoughts and beliefs of others?

e) Can you suggest any way round some of the ethical issues that are raised by others?

Interest Groups	Wellbeing (safety, welfare and health)	Autonomy (freedom and choice)	Justice (fairness)
Patients - people who are hoping that stem cell therapies will treat an illness, disease or injury.			
Scientists - people working in stem cell research, developing stem cell therapies to treat patients.			
Embryo - the source of embryonic stem cells for research.			
Society - issues for wider society, such as social priorities, research and medical priorities, and how money should be allocated.			

4.10 Learning points

Summary

- Cellular differentiation is the process by which a cell expresses certain genes to produce proteins characteristic for that type of cell. This allows a cell to carry out specialised functions.

- Meristems are regions of unspecialised cells in plants that can divide (self-renew) and/or differentiate.

- Stem cells are unspecialised cells in animals that can divide (self-renew) and/or differentiate.

- In the very early embryo, a group of cells, the inner cell mass, can differentiate into almost all body tissues and so are pluripotent.

- Tissue stem cells are involved in the growth, repair and renewal of the cells found in that tissue. They are multipotent.

- Tissue stem cells are multipotent as they can differentiate into all of the types of cell found in a particular tissue type. For example, blood stem cells located in bone marrow can give rise to all types of blood cell.

- Stem cell research provides information about how cell processes, such as cell growth, gene expression and gene regulation, occur.

- Stem cells are used in therapies to repair and replace damaged organs and tissues; more therapies are being developed to treat diseases such as diabetes.

- Research uses involve stem cells being used as model cells to study how diseases develop or being used for drug testing.

- There are major ethical issues surrounding stem cell use and research; for example, the use of embryonic stem cells involves the destruction of embryos which some people believe is akin to murder.

4.11 Extended response question

The activity which follows presents an extended response question similar to the style that you will encounter in the examination.

You should have a good understanding of stem cells before attempting the question.

You should give your completed answer to your teacher or tutor for marking, or try to mark it yourself using the suggested marking scheme.

Extended response question: Stem cells

Describe the differences between and similarities of embryonic stem cells and tissue stem cells. *(6 marks)*

4.12 Extension materials

The material in this section is not examinable. It includes information which will widen your appreciation of this section of work.

Extension materials: The Jacob-Monod hypothesis

Although bacterial gene control is not strictly needed for the exam, the simpler mechanisms in prokaryotes lead to better understanding of control in complex cells.

There are many different ways in which gene expression is controlled. The mechanisms of gene expression are very complicated and are not fully understood in higher organisms. However, a lot is known about gene expression in bacteria thanks to work carried out by two French scientists. In the 1950s, Francois Jacob and Jacques Monod, two scientists at the Pasteur Institute in Paris, performed a series of experiments and determined how the production of the enzyme β-galactosidase was controlled in the bacterium *Escherichia coli*. They later won the Nobel prize for their work.

E. coli uses glucose during respiration to release energy. It will always use glucose if it is present in the environment. If no glucose is available, then it will utilise the glucose found in other energy sources, such as lactose. However, it is only able to use the glucose in lactose once it has been separated from galactose. The enzyme β-galactosidase breaks down lactose into glucose and galactose. Jacob and Monod found that *E. coli* only produces β-galactosidase when lactose is present in the nutrient medium in which the bacteria are growing. If lactose is absent, then no β-galactosidase is produced.

The gene that controls β-galactosidase production is 'switched on' (expressed) in the presence of lactose and 'switched off' (not expressed) when lactose is absent. Enzyme induction occurs, meaning that the gene is only switched on when the enzyme it codes for is required. In other words, *E. coli* is able to regulate the expression of the genes needed for lactose metabolism.

The Jacob-Monod hypothesis suggests that the production of β-galactosidase is controlled by an operon. An operon is a region of DNA that contains an operator gene that controls the expression of a structural gene which make the proteins or enzyme. A regulator gene that is found further along the DNA strand codes for a repressor molecule that interacts with the operator gene, preventing the expression of the structural gene.

When lactose is absent, the regulator gene produces a repressor molecule. This repressor molecule binds to the operator gene and this means that the structural gene is 'switched off'.

When lactose is present, it binds to the repressor molecule which means the operator gene is free. The operator gene then 'switches on' the structural gene which produces the enzyme β-galactosidase. This enzyme breaks down lactose into glucose and galactose. When the lactose is used up, the repressor molecule binds to the operator gene and the structural gene is once again 'switched off'.

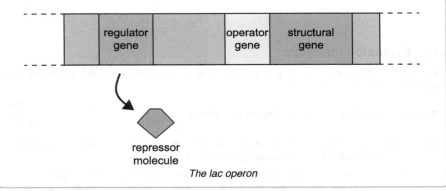

The lac operon

4.13 End of topic test

End of Topic 4 test Go online

Q5: Which of the following statements is **not** true of a plant meristem?

a) Growth can only occur at a meristem in a plant.
b) It contains unspecialised cells that differentiate.
c) It contains specialised cells that differentiate.
d) The roots and shoots have meristems.

Q6: The following diagram shows a cross section through a root.

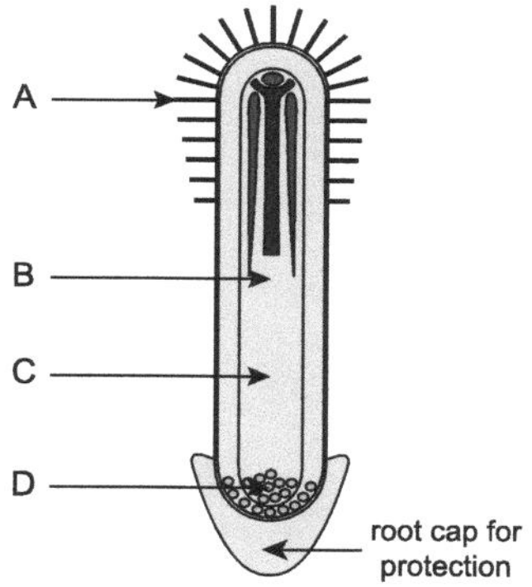

root cap for protection

Which letter indicates the meristem?

Q7: What is a stem cell?

Q8: State two properties of stem cells.

Q9: Name two sources of tissue (adult) stem cells.

Q10: Why do our bodies need stem cells?

Q11: Describe a current medical use of stem cells mentioned in this topic.

Q12: What is the main ethical consideration regarding the use of embryonic stem cells?

Unit 1 Topic 5

Structure of the genome

Contents

5.1 Introduction to the Genome 68
5.2 The genome 68
5.3 Learning points 70
5.4 Extension materials 71
5.5 End of topic test 72

Prerequisites

You should already know that:

- DNA carries the genetic information for making proteins.

Learning objective

By the end of this topic, you should be able to:

- understand the meaning of the term genome;
- describe that a section of DNA which produces a polypeptide is regarded as a gene;
- recognise that some DNA sequences code for proteins, while other sections do not;
- describe how some of the non-coding DNA has a role in controlling and regulating transcription;
- recognise that not all of the functions of non-coding DNA are as yet known or understood;
- describe the functions of RNA, including tRNA and rRNA.

5.1 Introduction to the Genome

The **genome** is the sum total of all the hereditary material within an organism. It is usually taken to be the complete complement of DNA, although in some viruses this could be DNA or RNA.

The majority of the genetic information is carried within the nuclear DNA (linear chromosomes), but other sources exist. In bacteria, the DNA is found as a circular strand, sometimes called a chromosome, but lacking the associated packaging proteins. Bacteria often contain plasmids, which are small circular sections of DNA.

In eukaryotes, organelles such as **chloroplasts** and **mitochondria** also contain circular sections of DNA. Mitochondrial DNA is of great importance in hereditary studies as it is only passed down the maternal line. More often, the term genome is used to refer to nuclear DNA only. The mitochondrial DNA may be referred to as the mitochondrial genome and the DNA of the chloroplast as the **plastome**.

The human genome is recognised as consisting of 3×10^9 nucleotides. These are found as approximately 20 - 25,000 genes, arranged on 22 autosomal chromosomes, and a pair of sex chromosomes, either two X chromosomes or an X and a Y chromosome. The study of the properties of genomes is referred to as **genomics**, compared to the study of single genes or groups of genes, which is **genetics**.

5.2 The genome

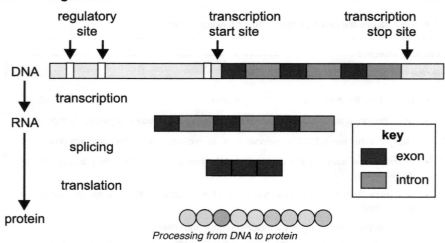

Processing from DNA to protein

An organism's genome is its genetic information encoded in its DNA. The genome contains many genes which carry instructions for making all of the proteins found in an organism. These regions of DNA are known as coding regions. The genome contains both coding and non-coding regions.

TOPIC 5. STRUCTURE OF THE GENOME

In fact, most of the genome is made up of non-coding sequences. These non-coding regions can have several functions:

- regulating transcription,
- transcription of RNA,
- no known function.

Some non-coding sections of DNA are used to regulate transcription. This means they can bind proteins which promote or prevent transcription of a gene. The diagram below illustrates how a sequence of DNA can regulate transcription of a gene.

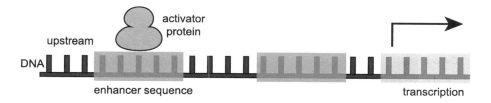

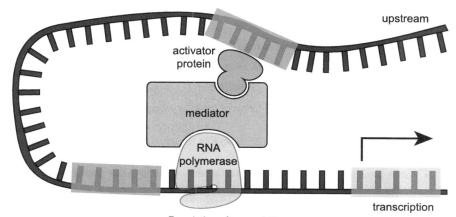

Regulation of transcription

Some sections of DNA are transcribed into RNA but are not translated, for example tRNA, which carries specific amino acids to the ribosome during translation, and rRNA, which together with protein forms the ribosome. Another type of RNA which is not translated is RNA fragments. These are small sections of RNA which are involved in splicing, and other processes such as post-transcriptional regulation of genes.

The function of large sections of the genome are still unknown. It was once referred to as junk DNA, but it is now widely acknowledged that it serves a purpose.

The genome: Question Go online

Q1: Complete the table of genome terms by matching the descriptions from the list with the processes.

Process	Description
Transcription	
Splicing	
Translating	

Descriptions:

- DNA copied to RNA
- Exons pass to ribosome where polypeptides are assembled
- Introns removed from pre-mRNA

5.3 Learning points

Summary

- The genome of an organism is its hereditary information encoded in DNA.
- Coding regions are sections of DNA which contain a gene.
- Much of the genome is non-coding in that it does not contain genes.
- Some regions of the genome contain regulatory sequences which control the transcription of genes.
- Other non-coding regions contain sequences for producing non-translated forms of RNA such as tRNA, rRNA and RNA fragments.
- While some sections of non-coding DNA assist in the control and regulation of gene expression, there are sections whose function remains unknown.

5.4 Extension materials

The material in this section is not examinable. It includes information which will widen your appreciation of this section of work.

Extension materials: The Human Genome Project

The Human Genome Project began in October 1990 and was completed in 2003. The project involved the discovery of all the estimated 20,000 - 25,000 human genes, making them available for further studies. The project also led to the discovery of the complete sequence of the 3 billion DNA sub-units (the bases in the human genome). In April 2003, the completion of the human DNA sequence coincided with the 50th anniversary of Crick and Watson's description of the DNA structure in 1953.

Only about 3 percent of the human genome is actually used as the set of instructions. These regions are called coding regions. At present, little is known functionally for most of the remaining 97 percent of the genome; these regions are called non-coding regions.

Single nucleotide polymorphisms, or SNPs (pronounced "snips"), are DNA sequence variations where a single nucleotide (A,T,C,or G) in the genome sequence is altered. For example, a SNP might change the DNA sequence ACGGCTCA to ATGGCTCA.

Remember that there are 20 different amino acids. DNA is made up of four nucleotides, so if each were to specify (or code for) a single amino acid, only four amino acids could be coded for. A two letter code would give 16 (4^2) possible arrangements - still not enough to code for all 20 amino acids. The shortest unit that can code for all amino acids is a triplet code or codon. A triplet code produces 64 (4^3) possibilities - more than enough!

A variation is considered a SNP when it occurs in at least 1% of the population. SNPs make up about 90% of all human genetic variation and can occur every 100 to 300 bases along the 3 billion base human genome. About 66% of SNPs involve the replacement of cytosine (C) with thymine (T). SNPs can occur in coding and non-coding regions of the genome. They can act as biological markers, helping scientists in locating genes that are associated with certain diseases.

SNPs have no effect on cell function; scientists believe SNP maps will help them identify the multiple genes associated with complex ailments such as cancer, diabetes, vascular disease, and some forms of mental illness.

Craig Venter's genome was published in 2007. His genome contains 4.1 million variations; 3.2 million were SNPs. The following year, James Watson's genome was published, costing about £8 million and taking only 4 months. In 2010, the first personal genome machine came onto the market. This machine can sequence an individual's genome in about 12 days at a cost of £6,000!

Although SNPs do not cause diseases, they can help determine the likelihood that someone will develop a particular illness. One of the genes associated with Alzheimer's disease, *apolipoprotein E* (or ApoE), is a good example of how SNPs affect disease development. Scientists believe that SNPs may help them to discover and catalogue the unique sets of changes involved in different types of cancers. They are confident that SNPs can play an important role in the different methods used in the treatment of cancer.

Scientists are trying to identify all of the different SNPs in the human genome. They are sequencing the genomes of a large number of people and then comparing the base sequences to discover SNPs. The sequence data is being stored in computers that can generate a single map of the human genome, containing all possible SNPs.

5.5 End of topic test

End of Topic 5 test — Go online

Q2: Complete the sentence using the words from list.

The _____ of an organism is its _____ information encoded in _____.

Word list: DNA; genome; hereditary

Q3: DNA sequences that code for proteins are:

a) cistrons
b) exons
c) genes
d) introns

Q4: All DNA codes for proteins. True or false?

Q5: Name two types of RNA.

Q6: The function of all DNA was worked out with the completion of the Human Genome Project.
True or false?

Q7: By working out the nucleotide sequence of a genome, it is possible to describe all the organism's genes.
True or false?

Unit 1 Topic 6

Mutations

Contents

6.1 Single gene mutations 75
    6.1.1 Types of mutation 75
    6.1.2 Effects of mutations 83
6.2 Chromosome structure mutations 86
6.3 The importance of mutations and gene duplication 89
6.4 Learning points 89
6.5 Extended response question 90
6.6 Extension materials 90
6.7 End of topic test 91

Prerequisites

You should already know that:

- DNA carries the genetic information for making proteins;
- a mutation is a random change to genetic material;
- mutations may be neutral, or confer an advantage or a disadvantage;
- mutations are spontaneous and are the only source of new alleles;
- environmental factors, such as radiation and chemicals, can increase the rate of mutation.

UNIT 1. DNA AND THE GENOME

Learning objective

By the end of this topic, you should be able to:

- understand and explain the term mutation;
- describe and explain single gene mutations and their consequences;
- describe the impact of mutations on evolution;
- describe and explain chromosome structure mutations and their consequences.

6.1 Single gene mutations

This section considers the types and effects of single gene mutations.

6.1.1 Types of mutation

Mutations are changes in the DNA that can result in no protein or an altered protein being synthesised. Single gene mutations involve changes to the number or the sequence of nucleotides within a single gene. There are three types of gene mutations: **substitution**, **insertion** and **deletion**.

Substitution mutation

A substitution mutation means that one nucleotide is substituted for another and an incorrect amino acid may be inserted into a protein. Usually these changes are minor, but they can cause major problems in some cases, i.e. sickle cell anaemia.

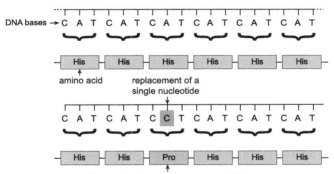

A substitution mutation

A substitution mutation within a protein coding gene may not always lead to a change in the amino acid sequence of the encoded protein. In this case, the mutation is described as silent. For example, the amino acid leucine might be encoded in a protein by the base triplet GAT, for which the corresponding mRNA codon would be CUA. If the T in the DNA base triplet mutates to C (to give GAC), the mRNA codon becomes CUG, and this still encodes leucine. Substitution mutations affect only one amino acid (if any) in the encoded protein.

An example of a disease caused by a substitution mutation is sickle-cell anaemia. The affected gene encodes beta-globin, a protein that forms part of haemoglobin. A GAG codon is changed to GUG. The result is that the amino acid valine is coded for instead of glutamic acid. In individuals affected by this disease, many of the red blood cells form a characteristic sickle shape and can get trapped in blood vessels. This can cause extensive tissue and organ damage. The life span of the red blood cells is considerably reduced and, because they cannot be replaced quickly enough, the individual develops anaemia.

76   UNIT 1. DNA AND THE GENOME

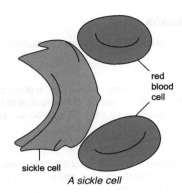

(a) DNA   TGA | GGA | **CTC** | CTC | TTC
     mRNA   ACU | CCU | **GAG** | GAG | AAG
     protein thr | pro | **glu** | glu | lys

T mutates to A

(b) DNA   TGA | GGA | **CAC** | CTC | TTC
     mRNA   ACU | CCU | **GUG** | GAG | AAG
     protein thr | pro | **val** | glu | lys

The sickle cell genetic mutation

Insertion mutation

Insertion mutations are caused by the addition of one or more nucleotides into a section of DNA. If one or two nucleotides are inserted into a protein coding gene, this can have drastic effects on the protein which is produced because all of the subsequent triplets are read incorrectly. The protein which is made is therefore likely to have many different amino acids and may not work at all.

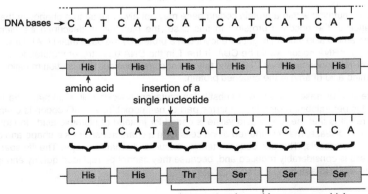

An insertion mutation

Deletion mutation

A deletion mutation refers to the removal of one or more nucleotides from the DNA. As with an insertion mutation, a deletion mutation alters the pattern of base triplets in the DNA. This means that deletions of one or two nucleotides are likely to cause drastic changes to a protein if they occur in a section of DNA containing a gene.

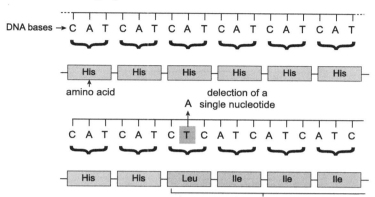

A deletion mutation

Gene mutations: Examples

The following provides examples of the three types of gene mutations.

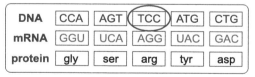

deletion mutation
The base at position 7 in the DNA (T) has been deleted.

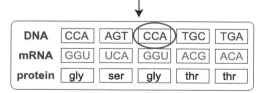

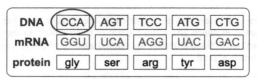

insertion mutation
A new base (G) has been inserted at position 2 in the DNA.

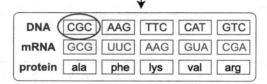

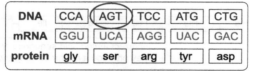

substitution mutation
The base at position 4 in the DNA has changed from A to T.

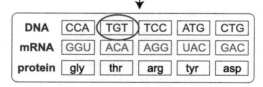

TOPIC 6. MUTATIONS

Gene mutations: Questions

Go online

Complete the sequences of proteins after mutation in each of the examples by referring to the genetic code table.

		Second letter					
		U	C	A	G		
First letter	U	phe	ser	tyr	cys	U	Third letter
	U	phe	ser	tyr	cys	C	
	U	leu	ser	stop	stop	A	
	U	leu	ser	stop	trp	G	
	C	leu	pro	his	arg	U	
	C	leu	pro	his	arg	C	
	C	leu	pro	gln	arg	A	
	C	leu	pro	gln	arg	G	
	A	ile	thr	asn	ser	U	
	A	ile	thr	asn	ser	C	
	A	ile	thr	lys	arg	A	
	A	met	thr	lys	arg	G	
	G	val	ala	asp	gly	U	
	G	val	ala	asp	gly	C	
	G	val	ala	glu	gly	A	
	G	val	ala	glu	gly	G	

Genetic code table

Q1:

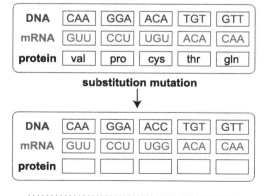

substitution mutation

....

Q2:

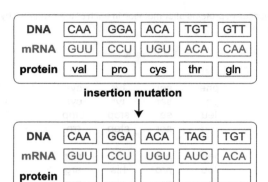

Q3:

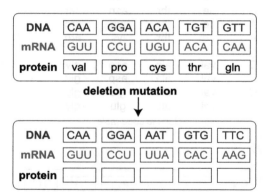

TOPIC 6. MUTATIONS

Complete the sequences of proteins after mutation in each of the examples by referring to the genetic code table and identify what kind of mutation has taken place.

		Second letter				
First letter		U	C	A	G	**Third letter**
	U	phe	ser	tyr	cys	U
	U	phe	ser	tyr	cys	C
	U	leu	ser	stop	stop	A
	U	leu	ser	stop	trp	G
	C	leu	pro	his	arg	U
	C	leu	pro	his	arg	C
	C	leu	pro	gln	arg	A
	C	leu	pro	gln	arg	G
	A	ile	thr	asn	ser	U
	A	ile	thr	asn	ser	C
	A	ile	thr	lys	arg	A
	A	met	thr	lys	arg	G
	G	val	ala	asp	gly	U
	G	val	ala	asp	gly	C
	G	val	ala	glu	gly	A
	G	val	ala	glu	gly	G

Genetic code table

Q4:

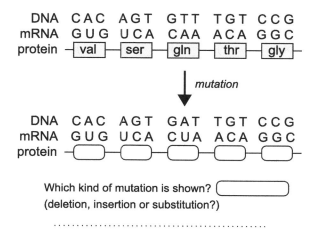

DNA CAC AGT GTT TGT CCG
mRNA GUG UCA CAA ACA GGC
protein —[val]—[ser]—[gln]—[thr]—[gly]—

↓ *mutation*

DNA CAC AGT GAT TGT CCG
mRNA GUG UCA CUA ACA GGC
protein —( )—( )—( )—( )—( )—

Which kind of mutation is shown? ⬚
(deletion, insertion or substitution?)

....

Q5:

```
DNA    CAC  AGT  GTT  TGT  CCG
mRNA   GUG  UCA  CAA  ACA  GGC
protein —[val]—[ser]—[gln]—[thr]—[gly]—
```

↓ *mutation*

```
DNA    CAC  TAG  TGA  TTG  TCC
mRNA   GUG  AUC  ACU  AAC  AGG
protein —[ ]—[ ]—[ ]—[ ]—[ ]—
```

Which kind of mutation is shown? [   ]
(deletion, insertion or substitution?)

....

Q6:

```
DNA    CAC  AGT  GTT  TGT  CCG
mRNA   GUG  UCA  CAA  ACA  GGC
protein —[val]—[ser]—[gln]—[thr]—[gly]—
```

↓ *mutation*

```
DNA    CAA  GTG  ATT  GTC  CGA
mRNA   GUU  CAC  UAA  CAG  GCU
protein —[ ]—[ ]—[ ]—[ ]—[ ]—
```

Which kind of mutation is shown? [   ]
(deletion, insertion or substitution?)

6.1.2 Effects of mutations

The effect of a mutation will depend on its type and location. A protein requires the correct sequence of amino acids to function properly. If the base sequence of a gene is disrupted, the amino acid sequence may be disrupted as well.

A substitution mutation occurs when one base is swapped for another. The effects of this type of mutation will vary depending on where they occur. Some effects of substitution mutations include:

- missense,
- nonsense,
- splice site mutations.

A missense mutation results in a single incorrect amino acid being inserted into a protein. The effect this altered amino acid has on the function of the protein will vary depending on its location and chemical properties. A nonsense mutation results in the code for an amino acid being changed to a stop codon. This can result in an abnormally short protein which may not function properly.

Splice-site mutations result in some introns being retained and/or some exons not being included in the mature transcript and may result in a non-functional protein. Splice site mutations such as these alter post-transcriptional processing.

Insertions and deletions usually have greater effects than substitutions, especially if 1 or 2 bases are inserted or deleted. Nucleotide insertions or deletions result in frame-shift mutations.

Remember, mRNA is read in groups of three (codons). If 1 or 2 DNA nucleotides are inserted or deleted, all the bases downstream are moved up or down from their place; this means the reading frame is altered. Frame-shift mutations cause all of the codons and all of the amino acids after the mutation to be changed. This has a major effect on the structure of the protein produced.

If the mutation occurs early in the sequence, then the overall effect is far greater than if it occurred later. In addition to coding for different amino acids, the 'stop' sequence will become misplaced which could result in the polypeptide being either too long or too short, but in any event greatly altered.

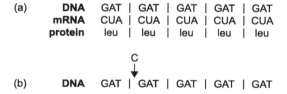

A frame shift mutation in DNA caused by insertion of one nucleotide

84   UNIT 1. DNA AND THE GENOME

Effects of gene mutations on amino acid sequences: Questions   Go online

A gene is a region of DNA which consists of a specific sequence of nucleotide bases arranged in triplets. Every amino acid is coded for by one or more of these triplets. Therefore, the sequence of bases determines the sequence in which amino acids are joined together to form a polypeptide or protein. Thus, a gene codes for a particular protein or polypeptide.

A gene mutation is a change in the sequence or type of nucleotide bases in a strand of DNA. This can lead to a change in the sequence of amino acids, and thus to a change in the protein which is synthesised in the ribosomes.

Sometimes a mutation causes only a minor change, perhaps affecting only one amino acid. This is known as a *point mutation* and the protein produced may be only slightly altered and still able to carry out its function.

Sometimes, however, a mutation can cause a major change affecting the coding for many amino acids. Such a mutation is known as a *frameshift mutation* and leads to a completely different protein being produced, which cannot carry out the required function.

Each of the four types of gene mutation are described in the following diagrams, with particular respect to changes in the amino acid sequences, which show the codons of part of a DNA strand and the amino acids which are coded for by them. Remember that the same amino acid can be coded for by more than one DNA triplet.

Q7: This is a deletion mutation. The first 'A' nucleotide in the original DNA strand has been removed (or deleted).

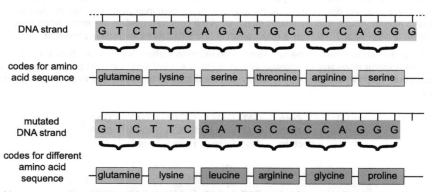

How many amino acids in this sequence have been changed? Is a deletion mutation a point or a frameshift mutation?

....

Q8: This is an insertion mutation. The 'T' nucleotide has been inserted between the 'C' and 'A' nucleotides in the original DNA strand.

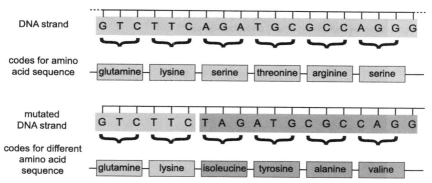

How many amino acids in this sequence have been changed? Is an insertion mutation a point or a frameshift mutation?

....

Q9: This is a substitution mutation. The 'T' nucleotide in the middle of the original DNA strand has been replaced (or substituted) by a 'G' nucleotide.

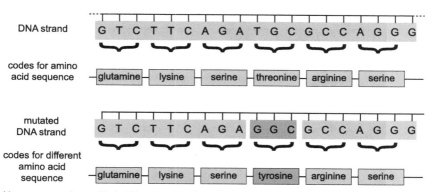

How many amino acids in this sequence have been changed? Is a substitution mutation a point or a frameshift mutation?

Q10: A deletion of three nucleotides from the middle of a gene may not lead to such a major change in the structure of the encoded protein as the deletion of one nucleotide. Why?

....

Q11: Sometimes, a mutation (even the change of one base) will lead to a protein being produced that is shorter than the normal protein. Why is this?

6.2 Chromosome structure mutations

Many types of mutations can arise through changes in chromosome structure. A change in chromosome structure starts when a chromosome breaks. The cell then attempts to repair the break, but in doing so may not restore the chromosome to its original structure. Changes in chromosome structure can be very large, with the result that many genes may be affected. Each type of chromosome mutation is briefly described below:

- **Translocation**: a section of a chromosome is added to another chromosome, not its homologous partner.
- **Deletion**: a section of a chromosome is removed.
- **Duplication**: a section of a chromosome is added from its homologous partner.
- **Inversion**: where a section of chromosome is reversed.

Chromosome mutation: A visual representation Go online

The following provides a visual representation of the changes that occur in chromosome structure during the three types of chromosome mutations.

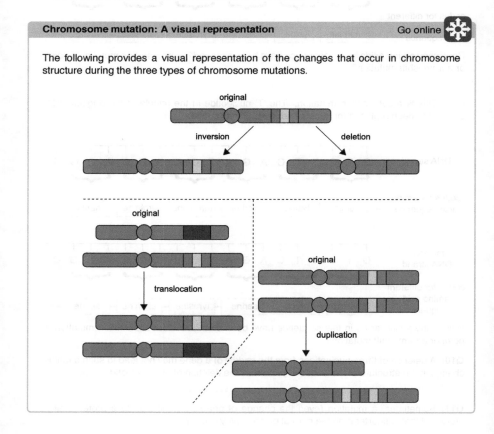

Differences between gene and chromosome mutation: Questions

Go online

A mutation is a change in the genetic material of an organism. Such changes can take place in the number or structure of chromosomes (known as chromosome mutations), or in the sequence of nucleotide bases in genes (known as gene mutations).

In each of the following examples, consider first of all whether it is a chromosome or a gene mutation, and then which particular type of mutation it is. Answer the questions by typing your answers in the boxes provided.

Q12:

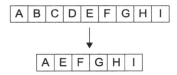

Is this a chromosome or a gene mutation and what is the name of this type of mutation?

....

Q13:

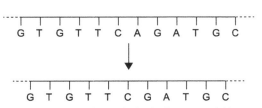

Is this a chromosome or a gene mutation and what is the name of this type of mutation?

....

Q14:

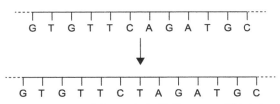

Is this a chromosome or a gene mutation and what is the name of this type of mutation?

Q15:

```
 |  |  |  |  |  |  |  |  |
 G  T  G  T  T  C  A  G  A  T  G  C
                    ↓
 |  |  |  |  |  |  |  |  |
 G  T  G  T  T  C  A  G  A  G  G  C
```

Is this a chromosome or a gene mutation and what is the name of this type of mutation?

Q16:

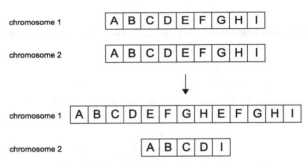

Is this a chromosome or a gene mutation and what is the name of this type of mutation (that occurs in chromosome 1)?

Q17:

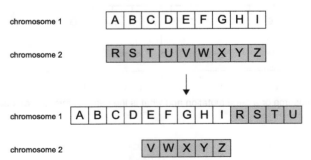

Is this a chromosome or a gene mutation and what is the name of this type of mutation (that occurs in chromosome 1)?

6.3 The importance of mutations and gene duplication

Mutations provide new variation. Mutations bring about new variation by the production of new alleles. Without mutations there would be no new variation.

Occasionally a mutation can result in the duplication of an entire gene. The second copy of the gene can become altered and provide new DNA sequences. This gene duplication is thought to be an important driving force in evolution.

6.4 Learning points

Summary

- A mutation is any change in the DNA sequence of the genome.
- Mutations can occur at random, but may also be induced by a number of agents.
- Mutations can occur at various levels. It could be change in a single nucleotide, allele, gene or chromosome.
- Point mutations are changes at the single nucleotide level, in that one nucleotide is substituted for another or a nucleotide is inserted or deleted from the DNA sequence.
- As a result of the change in the DNA sequence there may be a change in the amino acid sequence, and hence the structure of the final protein.
- Missense mutations result in one amino acid being changed for another. This may result in a non-functional protein or have little effect on the protein.
- Nonsense mutations result in a premature stop codon being produced which results in a shorter protein.
- Splice-site mutations result in some introns being retained and/or some exons not being included in the mature transcript.
- Frame-shift mutations cause all of the codons and all of the amino acids after the mutation to be changed. This has a major effect on the structure of the protein produced.
- As a result of mutations producing altered proteins, organisms may change or evolve.
- Mutations may occur at the chromosomal level, either within the chromosome or between chromosomes.
- The sequence of genes on a chromosome may be changed by duplication, deletion, translocation or inversion.
- Duplication is where a section of a chromosome is added from its homologous partner.
- Deletion is where a section of a chromosome is removed.
- Inversion is where a section of chromosome is reversed.
- Translocation is where a section of a chromosome is added to a chromosome, not its homologous partner.

> **Summary continued**
> - The substantial changes in chromosome mutations often make them lethal.
> - Duplication allows potential beneficial mutations to occur in a duplicated gene whilst the original gene can still be expressed to produce its protein.

6.5 Extended response question

The activity which follows presents an extended response question similar to the style that you will encounter in the examination.

You should have a good understanding of gene mutations before attempting the question.

You should give your completed answer to your teacher or tutor for marking, or try to mark it yourself using the suggested marking scheme.

> **Extended response question: Gene mutations**
>
> Describe gene mutations and outline some of their consequences. *(8 marks)*

6.6 Extension materials

The material in this section is not examinable. It includes information which will widen your appreciation of this section of work.

> **Extension materials: Lactose tolerance**
>
> In most parts of the world, the ability of humans to digest lactose declines rapidly after infancy. Lactose is a sugar that is found in both human breast milk and cow's milk. However, many adults in Northern Europe, and to some extent around the Mediterranean, retain the ability to digest lactose, and are thought to have developed this through a mutation followed by evolutionary adaption. It has been proposed that this ability may have arisen in response to living at 'high' latitude which, as a result of diminished sunlight, may lead to reduced levels of vitamin D. The vitamin is associated with calcium uptake and its absence could lead to bone malformation - rickets for example.
>
> By drinking fresh milk, which is high in calcium, or ingesting other dairy products, the problem of calcium deficiency could be overcome.

Through research and analysis, it has been suggested that the most probable explanation to the lactose tolerance is an evolutionary adaptation to millennia of milk drinking from domestic livestock. It has also been shown that the process of milking predated the evolution of lactose digestion. This would suggest that, as the lactose was present in the environment, there was an adaptation to exploit it.

6.7 End of topic test

End of Topic 6 test Go online

Q18: Which of the following correctly describes the characteristics of mutations?

a) They are non-random, frequent occurrences.
b) They are random, frequent occurrences.
c) They are non-random, infrequent occurrences.
d) They are random, infrequent occurrences.

....

Q19: What name is given to a mutation where one part of a chromosome becomes attached to another?

a) Deletion
b) Duplication
c) Inversion
d) Translocation

....

Q20: Which of the following chromosome mutations is most likely to be lethal?

a) Deletion
b) Duplication
c) Inversion
d) Translocation

Q21: The figure below shows part of the normal nucleotide sequence of a gene. N refers to an unknown nucleotide. The figure below also shows the effects of different types of mutations, indicated by A to C, on the nucleotide sequence.

Normal sequence	N N C A C G T A A C G T N N
A	N N C A C G T A A C C G T N
B	N N C A G T A A C G T N N N
C	N N C A C G A A A C G T N N

In the order A to C, what are the different types of mutations shown in the figure?

a) Substitution, insertion, deletion.
b) Insertion, substitution, deletion.
c) Insertion, deletion, substitution.
d) Substitution, deletion, insertion.

Q22: Which of the following gene mutations may result in a frameshift mutation?

a) Insertion or substitution.
b) Substitution or deletion.
c) Insertion or deletion.

Q23:

1. What word is used to describe a mutation which results in the addition of a premature stop codon?
2. What word is used to describe a mutation which results in the addition of an incorrect amino acid into a protein?

Unit 1 Topic 7

Evolution

Contents

7.1 Evolution 95
7.2 Gene transfer 96
7.3 Selection 96
    7.3.1 Natural selection 96
    7.3.2 Effects of Selection 100
7.4 Speciation 101
    7.4.1 Species 101
    7.4.2 Speciation 101
    7.4.3 Barriers to species 103
    7.4.4 Speciation mechanisms 104
7.5 Learning points 105
7.6 End of topic test 106

Prerequisites

You should already know that:

- a mutation is a random change to genetic material;
- mutations may be neutral, or confer an advantage or a disadvantage;
- mutations are spontaneous and are the only source of new alleles;
- an adaptation is an inherited characteristic that makes an organism well suited to survival in its environment/niche;
- variation within a population makes it possible for a population to evolve over time in response to changing environmental conditions;
- natural selection/survival of the fittest occurs when more offspring are produced than the environment can sustain, only the best adapted individuals survive to reproduce, passing on the genes that confer the selective advantage;
- speciation occurs after a population becomes isolated whereas natural selection follows a different path due to different conditions/selection pressures.

UNIT 1. DNA AND THE GENOME

Learning objective

By the end of this topic, you should be able to:

- understand that changes in organisms take place over a long period of time and are caused by mutations;
- describe how genetic information is passed vertically during sexual and asexual reproduction, from parent to offspring;
- describe how genetic material can be passed horizontally in prokaryotes, resulting in very rapid evolution;
- explain that selection is a non-random process that leads to the increased presence of a particular gene or genes in a population;
- describe the process of natural selection;
- understand that deleterious (harmful) sequences are selected against, and removed from the population;
- understand that speciation is the process of generating a new species;
- explain how speciation is influenced by allopatric and sympatric factors.

7.1 Evolution

Evolution describes the changes that occur to a **species** over time, leading to offspring that are better adapted to survive in their environment than the previous generation. For evolution to occur, there must be changes to the **gene pool** and hence changes to the frequency of genes. The gene pool refers to all of the different genes of a particular species.

The **allele frequency** refers to the frequency of any **allele** in the population. The allele frequency is sometimes called the gene frequency, but this can be misleading as the term is used to describe the frequency of alleles, not genes. Remember that a gene may have several different alleles. While the frequency of a gene may not change, the frequency of each allele can. For example, the gene for eye colour may remain at a constant level within a population, but the frequency of the allele for blue eyes may increase or decrease.

Changes in the allele frequency can occur by several different mechanisms, as described in the next table.

Mechanism	Definition
Mutation	Creates multiple alleles for many genes in the gene pool.
Gene migration	The movement of alleles between populations by individuals arriving from a different population and breeding. These individuals have a different gene pool and therefore introduce new alleles into the population.
Genetic drift	Tends to occur in small populations and describes the change in allele frequency due to a chance event. Small populations that are isolated from each other can vary greatly due to changes in allele frequencies.
Non-random mating	Does not change the frequency of the alleles, but increases the number of homozygous individuals. Inbreeding is the most common form.
Natural selection	The frequency of an allele increases in a population if it provides a selective advantage.
Chance	Changes to the allele frequency due to random loss. For example, an individual possessing a certain allele may die or fail to reproduce so that allele is lost from the population.

Factors affecting allele frequency

7.2 Gene transfer

Eukaryotes can reproduce by sexual or asexual reproduction. During sexual reproduction, the genetic material of two parents is combined to produce a new organism. During asexual reproduction, a new organism is produced from a single parent. Both sexual and asexual reproduction are examples of vertical gene transfer, a process by which genes are transferred from parent(s) to offspring.

In prokaryotes, reproduction is most frequently carried out asexually by a form of mitosis called binary fission. Again, this is an example of vertical gene transfer. However, there are occasions where prokaryotes can pass genetic material between themselves. The genetic material may be part of the single circular chromosome or a plasmid. This type of inheritance is called horizontal gene transfer because genes are passed between members of the same generation, not between parents and offspring. Viruses can also carry out horizontal gene transfer.

As with any other organism, prokaryotes are subjected to environmental pressures that sometimes cause mutations. Mutations can appear quite rapidly because prokaryotes exist in massive numbers. Once in the population, mutations can be passed between members with ease by horizontal gene transfer. Horizontal gene transfer allows prokaryotes to experience rapid evolutionary change.

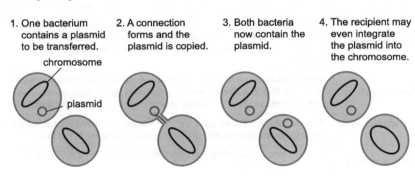

1. One bacterium contains a plasmid to be transferred.
2. A connection forms and the plasmid is copied.
3. Both bacteria now contain the plasmid.
4. The recipient may even integrate the plasmid into the chromosome.

7.3 Selection

This section considers natural selection and the effects of stabilising, disruptive and directional selection.

7.3.1 Natural selection

The environment surrounding all living organisms is never static and, as a result, they are constantly under pressure to respond to the changes in order to survive. Through sexual reproduction, the genetic material being passed between generations is subjected to constant change and rearrangement. As a result, after many generations the genome will be altered.

Natural selection is the mechanism by which evolution occurs. It is a process that selects the phenotypes that are best suited to the survival of a **species** in its particular environment. This means that the organisms which are most suited to their environment survive at the expense of those which are less well adapted. As the environment is changing continuously, natural selection is an ongoing process.

Working individually, Charles Darwin and Alfred Wallace suggested the same theory of evolution which they published in a joint paper in 1858. They proposed that natural selection was the mechanism by which evolution occurred. In his book *On The Origin of the Species*, Charles Darwin provided extensive evidence to support the theory of natural selection.

It is often assumed that evolution occurs over thousands, if not millions, of years, but in some cases it can be readily studied in organisms that have evolved over much shorter periods of time.

The theory of natural selection states that:

- In each generation, more offspring are produced than it is possible for the environment to support. Therefore, each individual in the offspring has to compete and struggle to survive so that it can reproduce and pass on its genes.
- Every individual in a population displays slightly different phenotypes. It is the individuals that possess characteristics that are most useful and better adapted to their environment that are most likely to survive. Less beneficial phenotypes are gradually removed from the general population as individuals displaying these characteristics have a reduced survival rate and a reduced chance of reproducing.
- This process continues over many generations, increasing the numbers of individuals displaying the advantageous characteristics for that environment so that they dominate the population. In this way the phenotypes that are beneficial to the organisms in their particular environment are selected and preserved within the **species**.
- The ability of an individual to reach adulthood and reproduce is described as its fitness. The "fitter" an organism is, the more likely it is to survive and produce offspring that go on to reproduce.

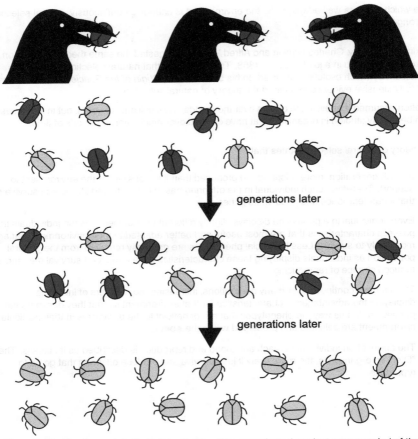

Natural selection in a nutshell: dark bodied beetles are selected against over a period of time whereas light bodied beetles flourish

TOPIC 7. EVOLUTION

Natural selection: Questions

Q1: Place the stages into the correct places to complete the diagram of the key stages that occur during natural selection:

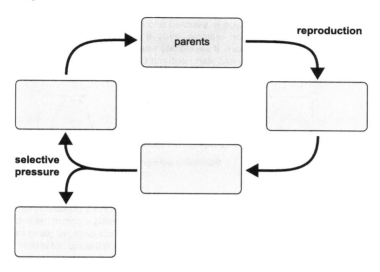

Q2: Which of the following statements is **not** true of evolution by natural selection?

a) Natural selection is an ongoing process.
b) The frequency of beneficial genes increases within a population.
c) All individuals in a population display the same phenotypes.
d) High levels of competition exist between individuals.

....

Q3: What does fitness of an individual describe in terms of evolution?

a) The number of times it mates during its lifetime.
b) The reproductive success it experiences during its lifetime.
c) The length of time it lives for.
d) How well it adapts to its environment.

7.3.2 Effects of Selection

Changes in phenotype frequency can occur as a result of stabilising, directional and disruptive selection.

1. *Stabilising Selection*
   In this case, the average phenotype is selected and the extremes survive much less well, possibly even disappearing. As an example, birds in a particular environment may have a range of colourings from light to dark. If the climate were to change to dull and overcast, then the white and black individuals would stand out and become prey to predators. The result would be an increase in grey birds because their grey (average) colour was selected for.

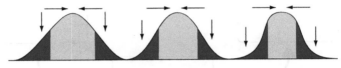

Stabilising selection

2. *Disruptive Selection*
   In this case, it is the extreme values of phenotypes that are chosen and those with average fitness are selected against. As an example, assume the same bird population as before, but now the climate changes and becomes colder with snow persisting in part of the habitat. White birds will be well hidden from predators in the snow and the black birds will blend into the dark background below the snowline. However, the grey individuals will stand out in both conditions and will thus be susceptible to predation. Now it is the extreme values or phenotypes that are selected for.

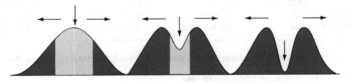

Disruptive selection

3. *Directional Selection*
   In this final case, one extreme value or phenotype is selected over both the average and the other extreme value. Based on the same bird population again, let us assume that the snow has gone and left a dark earth-scape. Now, both white and grey varieties will stand out and become victims of predation. The dark phenotype is selected for and the numbers of these birds rise as a result.

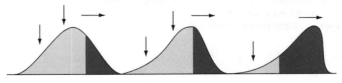

Directional selection

TOPIC 7. EVOLUTION

7.4 Speciation

This section considers the concept of species, the process leading to the formation of new species and its mechanisms, the barriers that can affect how species develop and the concept of hybrids.

7.4.1 Species

A **species** is described as a population of organisms that have the same characteristics and are capable of interbreeding to produce fertile offspring. For example, lions, tigers and jaguars all possess similar traits, such as body shape, facial and paw structure, and their ability to roar, but they are unable to mate with each other to produce fertile offspring, making them three distinct species.

Each member of the species has the same number of chromosomes and the same **gene pool**. As the gene pool is comprised of the sum of all of the different genes of a particular species, it follows that if there are changes to the gene pool, then evolution will occur.

Due to organisms moving and breeding with different populations of the same species, genes are continually moving between populations. When this stops, and populations become isolated, different species may emerge.

Species: Questions   Go online

Q4: What is a population?

....

Q5: What is the difference between the gene pool and allele frequency?

7.4.2 Speciation

Speciation is the generation of new biological species by evolution as a result of isolation, mutation and selection. Populations of an existing **species** can become isolated from each other, with the result that their gene pools diverge. The isolated populations experience different selection pressures and adapt to their particular niche, developing different characteristics. Individuals from the different populations will eventually no longer be able to breed with each other. A new species will then have been formed.

When interbreeding populations become separated from each other the flow of **alleles** between them is prevented. This means that the **gene pool** of each sub-population is no longer influenced by the gene pools of other sub-populations of the same species.

Within a population, any mutations that arise which are beneficial to the population are favoured by **natural selection**. In separated populations, new alleles may be introduced that cause the sub-populations to evolve in a slightly different way, eventually leading to the creation of new species.

© HERIOT-WATT UNIVERSITY

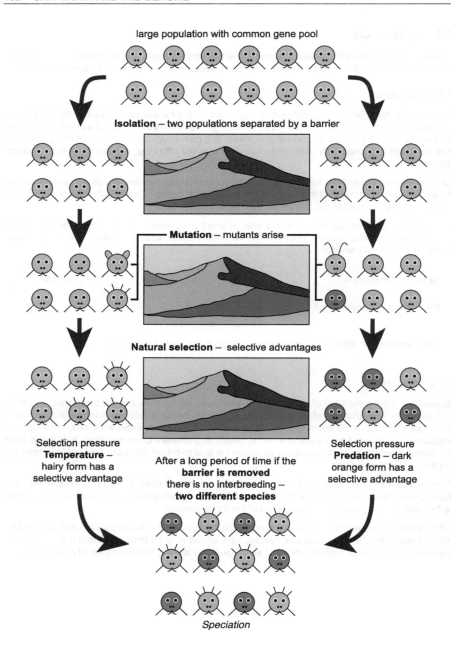

TOPIC 7. EVOLUTION

7.4.3 Barriers to species

There are several types of barriers which can bring about speciation:

- *geographical barriers*: these include mountains, deserts, oceans and rivers that physically separate organisms and prevent populations from interbreeding. Geographical isolation may also occur if a habitat is lost, such as the destruction of a forest to form an arid landscape or a river drying up.

- *ecological barriers*: factors such as temperature, pH, salinity, humidity and altitude also act to separate populations. For example, many species have evolved to inhabit regions of different pH or salinity.

- *behavioural barriers*: if individuals in a population become fertile at different times of the year, their sexual organs change, or their courtship behaviour is different or unattractive, then the individuals cannot mate.

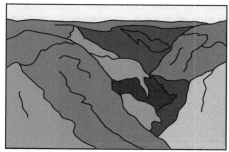
1. This canyon has been isolated by mountains.

2. Different terrains on neighbouring mountains.

3. Water creates a difficult barrier to cross.

4. Deserts are hostile environments.

Examples of geographical barriers to species

7.4.4 Speciation mechanisms

Allopatric speciation occurs due to populations becoming physically separated; it is brought about by geographical barriers. After separation, the now isolated populations may be subjected to different selective pressures or develop and maintain mutations which benefit that group. If, for any reason, the barriers are removed and the populations can freely intermingle but still cannot reproduce and produce fertile offspring, then speciation will have occurred and a new species will have been formed.

Sympatric speciation is a form of speciation where two species arise within the same habitat. For this to happen, other isolating mechanisms must be at work. Sympatric speciation occurs as a result of behavioural or ecological barriers.

Sympatric speciation is much more common in plants compared to animals. If parent plants produce offspring that are **polyploids**, this means that while these plants remain in the same habitat, they are now incompatible for breeding.

In a rare example of animal sympatric speciation, two groups of *Orcinus orca* (killer whale) live in the same habitat in the northeast Pacific ocean. Of these two groups, one is 'resident' and the other is 'transient'. Studies show that they stay apart from each other and do not interbreed; they have different diets, vocal behaviour and social structures. Although the two groups of whales currently belong to the same species, if this situation continues, speciation may occur in the future.

7.5 Learning points

Summary

- Evolution is the changes in organisms over generations as a result of genomic variations.
- Inheritance usually involves the passing of genetic material from parents to offspring. This is vertical inheritance.
- The passing of genetic material is brought about by sexual or asexual reproduction.
- Prokaryotes usually reproduce by binary fission, a form of vertical inheritance.
- Prokaryotes and viruses can also exchange genetic material between members of the same generation. This is horizontal inheritance.
- Horizontal inheritance frequently involves the exchange of whole or parts of a plasmid.
- Plasmid exchange will bring about rapid evolutionary change.
- Natural selection is the non-random increase in frequency of DNA sequences that increase survival and the non-random reduction in the frequency of deleterious sequences.
- Changes in phenotype frequency can be as a result of stabilising, directional and disruptive selection:
  - in stabilising selection, an average phenotype is selected for and extremes of the phenotype range are selected against;
  - in directional selection, one extreme of the phenotype range is selected for;
  - in disruptive selection, two or more phenotypes are selected for.
- A species is a group of organisms capable of interbreeding and producing fertile offspring, and which does not normally breed with other groups.
- Speciation is the generation of new biological species by evolution as a result of isolation, mutation and selection.
- Allopatric speciation occurs when populations become physically separated. This may be due to geographical barriers such as oceans or mountain ranges.
- Sympatric speciation is brought about by behavioural or ecological barriers. The process will usually occur within the same habitat.
- After a barrier separates a population, different mutations may arise within each subpopulation; natural selection will act differently on each group depending on the selection pressures present and, eventually, the two subpopulations become separate species.

7.6 End of topic test

End of Topic 7 test — Go online

Q6: Evolution can be described as a change in a _____ over time, and is driven by _____.

Q7: Inheritance can be described as the passing of _____ between generations.

Q8: The exchange of plasmids between bacteria is an example of _____ inheritance.

Q9: *Population change*

The Galapagos tortoises are very much at risk of extinction. The following table and graph show the rise in the human population in the Galapagos Islands over the last 100 years.

Year	Population size
1900	500
1920	500
1940	600
1960	2500
1980	4000
2000	20000

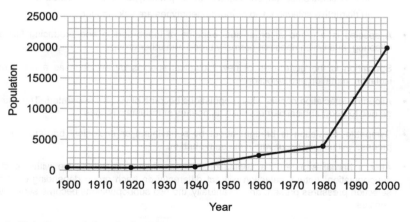

i. State the population size in 1990.
ii. By what percentage did the human population increase between 1980 and 2000?

Q10: Identify which graph represents stabilising selection, directional selection and disruptive selection.

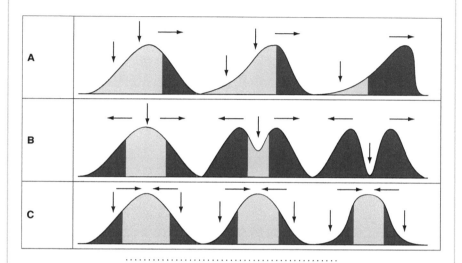

Q11: Match each of the following characteristics with one of the three selection types (stabilising selection, directional selection and disruptive selection).
   i. Average phenotype selected.
   ii. Extreme phenotype selected.
   iii. One extreme selected instead of the other one or the average.

Q12: Speciation is characterised by changes in gene frequency.
True or False?

Q13: In what way does sympatric speciation differ from allopatric speciation?

Unit 1 Topic 8

Genomics

Contents

- 8.1 Genomic sequencing 111
- 8.2 Phylogenetics 112
- 8.3 Comparative genomics 114
- 8.4 Personal genomics 115
- 8.5 Learning points 116
- 8.6 Extension materials 117
- 8.7 End of topic test 120

Prerequisites

You should already know that:

- the four bases A, T, G and C make up the genetic code;
- a mutation is a random change to genetic material;
- mutations may be neutral, or confer an advantage or a disadvantage.

Learning objective

By the end of this topic, you should be able to:

- understand that the complete nucleotide sequence for either a gene or a complete genome can be sequenced, requiring the use of bioinformatics;
- understand that, by using sequence data, it is possible to trace the evolution of organisms, how they are related to each other, and when they may have diverged from each other;
- understand that, by using sequence data in conjunction with the fossil record, the sequence of evolution can be established;
- understand that many genomes have been sequenced, including the human genome and other important species; comparisons of genomes show that large areas are conserved;
- understand that an individual's genome can be analysed to assess the likelihood of disease and, if necessary, design tailored treatment.

TOPIC 8. GENOMICS

8.1 Genomic sequencing

DNA sequencing is the process of determining the order of nucleotides in a section of DNA. It is now possible to determine the sequence of nucleotides in relatively small sections, i.e. a gene, or very large sections, i.e. a complete genome.

The process by which sequencing can be achieved involves several techniques, and it is only relatively recently that two fields, in particular, have become sufficiently advanced to be incorporated together. PCR and related techniques now produce nucleic acids in sufficient quantity and purity, and computer science and statistical analysis can now cope with the quantity of evidence to rationalise the data into a comprehensible form. The use of computers and statistical analysis is known as **bioinformatics**.

Sequencing techniques

The development of sequencing techniques began sometime after the development and refinement of the Crick and Watson model of DNA.

By the early 1970s, recombinant DNA technology was becoming established. From this, defined fragments of DNA could be generated, as opposed to samples from bacteriophage or viruses. By 1977, the first complete DNA **genome** had been established. At the time, two teams were establishing techniques of sequencing.

Early attempts at sequencing were developed by Sanger and Coulson in 1975, called the plus-minus method. This was overtaken by Maxam and Gilbert in 1977, by a method based on chemical modification of DNA followed by cleavage at specific bases and, while accurate, it proved difficult to scale up and involved extensive use of toxic chemicals.

At about the same time, Sanger and his team developed the so-called "Chain Termination" method; this, and its subsequent developments, has become the technique favoured by many. Originally, radioactive materials were used which posed a hazard, but they were replaced first with UV detection and then fluorescent dyes.

8.2 Phylogenetics

Phylogenetics has been described as the field of biology that deals with identifying and understanding the relationship between the different kinds of life on Earth, others use the term 'evolutionary relatedness'. It has become an essential tool in tracing the evolutionary tree of life as first proposed by Darwin.

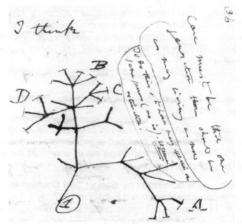

A diagram and note written by Charles Darwin

Originally, the relationship between organisms was traced by comparison of physical characteristics, embryology and examination of fossil records. Some of this is still in use but, more recently, advances in DNA sequencing have become much more prevalent. These days, sequence data and fossil evidence are both used to determine the main sequence of events in the evolution of life. Based on this relatively modern approach, life is now classified based on the three domains using a system first proposed by Carl Woese in 1977.

TOPIC 8. GENOMICS

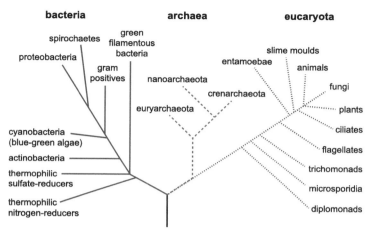

Three domains in the tree of life

It has been over 150 years since Darwin first put forward his idea that all life is related, and it is now widely accepted that all life originated in the sea some 3000 million years ago. The first cells developed and replicated themselves in two ways. Some formed into chains, which gave rise to algae and plant life, while others developed into hollow balls, which became sponges and the origins of animal life.

These groups grew and differentiated until about 450 million years ago with the development of land animals. As time progressed, greater divergence arose. Several mass extinction events eliminated many types of organism, but also gave those surviving the opportunity to flourish.

Over long periods of time (millions of years), mutations built up at a steady rate in any section of DNA. If the rate could be shown to be reliable, it could be used as a mechanism to estimate the time between mutations, i.e. a clock. This, in turn, can be used to estimate both where and when organisms diverged.

Molecular clocks are used to show when species diverged during evolution. They assume a constant mutation rate and show differences in DNA sequences or amino acid sequences. Therefore, differences in sequence data between species indicate the time of divergence from a common ancestor.

However, rates of molecular evolution can vary between organisms and so molecular clocks have to be calibrated. To do this, it is necessary to know the absolute age of some evolutionary divergence which can usually be determined from the fossil record.

Phylogenetics: Question   Go online

Q1: Complete the paragraph using the words from the list.

All life forms are now described as belonging to one of three _____. This is largely based on a comparison of _____. The three main groups are _____, _____ and _____. Phylogenetic clocks need to be calibrated by using _____.

Word list: archaea, bacteria, DNA, domains, eucaryota, fossil records.

8.3 Comparative genomics

The **genome** is the sum total of all the hereditary material within an organism. **Genomics** is the science of interpreting genes: the study of an organism's genome using information systems, databases and computerised research tools.

Many genomes have been sequenced, particularly those of disease-causing organisms, pest species and species that are important model organisms for research. The following table shows the size of genomes in some organisms.

Organism	Estimated size (base pairs)	Chromosome number	Estimated gene number
Human (*Homo sapiens*)	3 billion	46	ca. 25,000
Mouse (*Mus musculus*)	2.9 billion	40	ca. 25,000
Fruit fly (*Drosophila melanogaster*)	165 million	8	13,000
Plant (*Arabidopsis thalania*)	157 million	10	25,000
Roundworm (*Caenorhabditis elegans*)	97 million	12	19,000
Yeast (*Saccharomyces cerevisiae*)	12 million	32	6,000
Bacteria (*Escherichia coli*)	4.6 million	1	3,200

Comparison of some genomes

Comparative genomics is the process whereby the genomes of different species are compared. When comparing genomes from different species, scientists noted that many parts of the genome are highly conserved. This means that some sections of DNA are identical (or almost identical) between different species. These conserved regions of DNA are useful in determining evolutionary relationships.

One of the earlier organisms to be sequenced was the puffer fish. The attractions of the puffer fish, or fugu, are that it has one of the most compact genomes of all vertebrates. Roughly speaking, it contains a similar number of genes to humans, but they are contained in only 400 megabases compared to the 3.1 gigabases of the human genome.

By comparing the fugu genome to the human genome, it is possible to establish common functional elements of genes and regulatory sequences. By contrast, the non-functioning genes show where evolutionary divergence has occurred from a common ancestor, approximately 450 million years ago.

8.4 Personal genomics

One of the aims of **genomics** is to explain why some individuals are susceptible to disease while others seem unaffected. By understanding the interaction between genes and the environment, it may be possible to prevent the onset of some of these complex diseases in individuals.

Personal genomics is the sequencing and analysis of an individual's **genome**. Once an individual genotype (or part of it) is known it is compared to references in the published literature. From this, any mutations or sequences likely to give rise to disease can be identified. This is now referred to as predictive medicine, which in turn can lead to the use of an appropriate drug treatment if required, a process know as pharmacogenetics.

Key to personal genomics has been cost, which has been falling rapidly. When the first genome was sequenced, it cost in the region of three billion dollars; this genome was a composite of several individuals. Currently (May 2018), individuals can have their own genome (or sections of it) sequenced for £700 and completed within days or hours. However, after sequencing there must follow an analysis which may incur further costs and time.

As a result of advances in this field, a question of ethics has also arisen. Insurance companies, banks and others may decline services or increase premiums as a result of finding less desirable traits, e.g. Alzheimer's or other degenerative diseases. This has been termed genetic discrimination. As yet, regulations in this and associated fields are not clearly laid out.

Personal genomics could bring about greater understanding of the varying effects of drugs between different individuals. One example is the group of genes responsible for drug (and other metabolites) metabolism - cytochrome P450 (CYP) genes. Depending on which alleles have been inherited, an individual may be described as an extensive metaboliser (that is to say, normal), and can successfully metabolise certain compounds. Others, however, may be found to be intermediate or poor. As a consequence, patients may find that the drug of choice may be ineffective or cause severe adverse reactions. Clearly, it would be of significant advantage to have this information prior to treatment. It could save the patient from potential danger, and save the often considerable cost of medication.

8.5 Learning points

Summary

- It is possible to determine the sequence of bases in any section of DNA. It could be a relatively short sequence, for example a gene, up to and including a whole genome.
- To elucidate a DNA sequence, several different scientific disciplines are required.
    - First, DNA needs to be sequenced, which could be achieved by a variety of methods.
    - Second, the data collected must be analysed; this is achieved through mathematical, statistical and computing sciences (bioinformatics).
- From the DNA sequencing of different species' genomes, phylogenetic trees are formed. These show the relationship between organisms and their common ancestors.
- These comparisons are made by comparing DNA sequences and looking for mutations within comparable regions of a genome.
- Knowing the rate of mutation, it is possible to pinpoint the time at which divergence occurred.
- Mutation rates can be calibrated by cross reference to the fossil record.
- It is on this basis that classification of organisms is now based on the Three Domain system.
- Based on sequencing and fossil records, the path of evolution can be supported from the emergence of cells through prokaryotes, eukaryotes to higher plants and animals.
- The genomes of many organisms have been completed, and many more are being undertaken.
- Other than the human genome, organisms that cause disease in man, domestic animals and food crops have been sequenced. In addition, some organisms have been sequenced as they have proved to be most useful for comparison and modelling during research.
- It has become apparent when comparing genomes that sections are highly conserved, or similar.
- By studying an individual's genome, it is possible to determine errors or gather evidence to support the likelihood of ailments. Some examples include the predisposition to cancer, mental illness or drug dependency.
- It may be possible, armed with this knowledge, to treat or alleviate symptoms. Personalised medical care of this nature is called pharmocogenetics.

8.6 Extension materials

The material in this section is not examinable. It includes information which will widen your appreciation of this section of work.

Extension materials: Chain termination DNA sequencing method

The key principal of the chain termination DNA sequencing method is the use of dideoxynucleotide triphosphates (ddNTPs) as chain terminators.

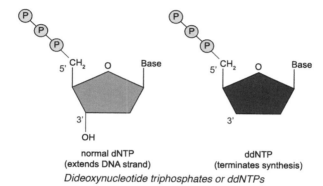

normal dNTP
(extends DNA strand)

ddNTP
(terminates synthesis)

Dideoxynucleotide triphosphates or ddNTPs

A single strand of DNA is required. This is called the DNA primer. Also needed are DNA polymerase, normal deoxynuclotide triphophates (dNTPs) and modified dideoxynucleotides (ddNTPs). These ddNTPs terminate DNA elongation and may be either radioactively tagged or fluorescently labelled. Labelling in this way allows for detection in automated machines.

The DNA sample is divided into four sequencing reaction vessels, containing normal deoxynucleotides and DNA polymerase. Then, each of the four separate vessels has one of the modified ddNTPs added to it. These ddNTPs stop chain elongation as they are lacking a 3'-OH group and so cannot form a phosphodiester bond to the next nucleotide. As a result, fragments of DNA of different lengths form.

The fragments are denatured and separated by gel electrophoresis. Each reaction is run in a separate lane. The bands of DNA are visualised depending on the marker chosen.

When using autoradiography, X-ray film may be used. If using fluorescent dye, a laser detector is used. Dark bands or specific colours dictate the position of the ddNTP at the end of a DNA strand. As each band appears, it indicates the sequence of nucleotides in the chain.

The use of fluorescent labelled ddNTPs and primers has allowed automated systems to be set up and has greatly increased the speed at which DNA sequencing can be carried out.

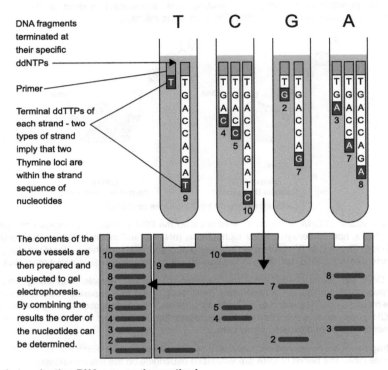

Chain termination DNA sequencing method

Following the successful completion of the Human Genome Project in 2003, one of the next goals was to develop the $1000 **genome**. That is, an individual's full genome to be described for less than $1000 in a matter of days.

The method has a few drawbacks, typically that some bases at either end of the sequence may be misread, but software can estimate these gaps.

Despite refinement and automation, faster methods have been sought. This is called high-throughput sequencing, or next generation sequencing, and utilises sequencing which runs many parallel processes producing vast numbers of sequences at a time.

Many of these techniques use emulsion PCR and nanopore sensing.

A technology that was announced in February 2012 suggests a significant advance towards the $1000 genome. It is based on nanopore technology and is designed to be capable of reading a single strand of DNA directly. As the DNA passes through the nanopore, it excites an electric current that is particular for each nucleotide. The advantage of this system is that it can measure single molecules directly without the need for amplification, labelling or optical reading instruments.

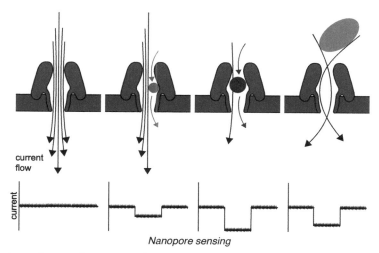

Nanopore sensing

The above diagram shows a protein nanopore set in an electrically resistant membrane bilayer. An ionic current is passed through the nanopore by setting a voltage across this membrane.

What are the benefits of using nanopores to sequence DNA?

In contrast to current sequencing technologies, nanopores can measure single molecules directly, without the need for nucleic acid amplification, fluorescent/chemical labelling or optical instrumentation.

It is claimed that using a series of these 'chips', or nodes, that a human genome could be sequenced in 20 minutes for approximately $1500 at today's prices. Smaller versions with a more limited capacity can be inserted into the USB of a laptop at a cost of $900.

8.7 End of topic test

End of Topic 8 test — Go online

Q2: Determining the order of nucleotide bases is known as _____.

Q3: What two techniques are combined to perform bioinformatics?

Q4: As a result of sequencing, phylogenetic trees can be formed. What is their purpose?

Q5: As well as sequence data, what other information is required to determine the main sequence of events in the evolution of life?

Q6: The _____ is the sum total of an organism's DNA.

Q7: Does the quality of DNA in an organism tell you the number of genes?

Q8: Many genes are found to be highly conserved across species. What does this mean?

Q9: What advantages might be gained from knowing an individual's genome?

Q10: What name is given to the field of medicine that aims to use knowledge of patients' genomes to design appropriate courses of medicines?

Unit 1 Topic 9
End of unit test

End of Unit 1 test

Go online

Structure of DNA

The diagram below represents the four different nucleotides found in DNA.

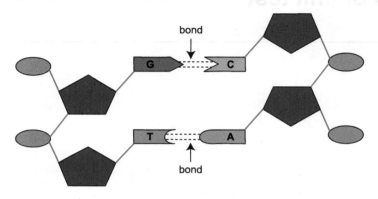

Q1: Name the type of bond which links the base on one DNA strand to the base on the opposite strand.

....

Q2: A section of a DNA molecule contains 8000 bases. Of these, 2800 are thymine. What is the percentage of cytosine bases in the molecule?

....

Q3: What is the name of the sugar present in DNA?

....

Q4: The diagram below shows a DNA nucleotide.

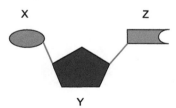

Identify the parts labelled X, Y and Z in the diagram.

Replication of DNA

The following diagram represents DNA replication.

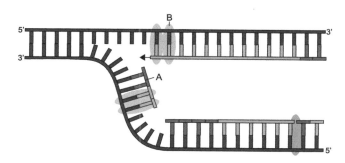

Q5: Name structure A.

....

Q6: Name one essential requirement for DNA replication that is not shown above.

....

Q7: Name enzyme B.

....

Q8: The polymerase chain (PCR) can be used to obtain many copies of a particular gene. Complete the following sentence.
The _____ bonds of DNA are separated by _____ during PCR.

....

Q9: Primers are required for the process of PCR. Which of the following statements is false?

a) Primers mark the start and end of sequence to be copied.
b) Primers prevents the strands re-joining.
c) Primers unwind double helix.
d) Primers allow DNA polymerase to attach.

....

Q10: Starting with a single molecule of DNA, the polymerase chain reaction goes through three complete cycles. How many molecules of DNA would be produced?

a) 4
b) 8
c) 16
d) 32

Gene expression

The following diagram shows the stages of protein synthesis.

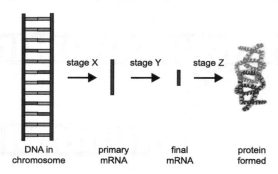

Q11: Name stage X.

....

Q12: Name the exact location in the cell where stage X occurs.

....

Q13: Name stage Z.

....

Q14: Name the exact location in the cell where stage Z occurs.

....

Q15: Which enzyme catalyses stage X of this process?

....

Q16: Explain why a primary mRNA molecule is longer than a mature mRNA molecule.

....

Q17: The mRNA codon for the amino acid methionine is AUG. What is the base sequence of the corresponding anti-codon?

....

Q18: What name is given to the chemical bonds that join amino acids together in a protein?

....

Q19: Name the bonds which hold polypeptide chains together in a protein.

Cellular differentiation

There are hundreds of cell types in the human body which originate from stem cells in the early embryo.

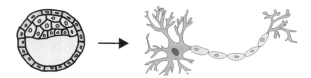

Q20: What are stem cells?

....

Q21: A well known drug company has developed a drug for the treatment of the symptoms of a particular inherited disease. However, before they can proceed to clinical trials using volunteers, the company has decided to do additional tests in the laboratory using stem cells. Suggest an ethical consideration which may influence their decision in using stem cells.

....

Q22: What is the difference between embryonic and tissue stem cells?

....

Q23: Meristems are regions of _____ cells in plants that are capable of cell _____.

Structure of the genome

Q24: What is a gene?

....

Q25: Name a non-translated form of RNA.

Mutations

Q26: Listed below are three types of mutation.

a) Deletion
b) Substitution
c) Insertion

Which mutation(s) affect only one amino acid in the protein produced?

....

Q27: What word is used to describe a mutation which results in the addition of an incorrect amino acid into a protein?

....

Q28: The sequences below show a type of single gene mutation.

   Original sequence: AATTGCTATG
   Mutated sequence: AATGGCTATG

Name the type of mutation shown by the sequences.

....

Q29: The following diagram shows a type of chromosome mutation.

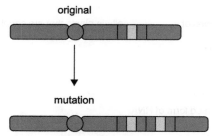

What type of mutation is shown in the diagram?

1. Deletion
2. Duplication
3. Inversion
4. Translocation

Evolution

Q30: Bacteria can on occasions pass genetic material between themselves. This may be a section of DNA called a _____.

....

Q31: This is representative of _____ genetic transfer.

....

Q32: Which of the following organisms are capable of vertical gene transfer?

a) Eukaryotes
b) Prokaryotes
c) Prokaryotes and eukaryotes

....

Q33: Complete the following sentences using the words from the list.

- _____ selection is where the average phenotype is most successful for a particular habitat.
- _____ selection is characterised by the extreme versions of a phenotype being selected.
- _____ selection is characterised by the selection of one extreme phenotype at the exclusion of all others.

Word list: Directional, Disruptive, Stabilising.

....

Q34: What is a species?

....

Q35: What type of barrier is involved in allopatric speciation?

....

Q36: Many species of cichlid fish are found in Lake Milawi. They are thought to have come from a common ancestor, but cannot interbreed as they fail to recognise each other's courtship patterns. Of the three species, **A** filter feeds on microbes in the water, **B** scrapes algae from rock surfaces and **C** crushes snail shells to extract the meat.

From the information above, why would **A**, **B** and **C** be considered different species?

....

Q37: From the information above, what is this type of speciation described as?

Genomics

Q38: Life can be divided into three domains. What process is used to produce evidence to support this claim?

....

Q39: Primates, including orangutans, gorilla, chimpanzee and humans, have a common ancestor from 35 million years ago. Humans and chimps diverged from each other 5 million years ago, gorillas 5 million years before that and orangutans 9 million years before that. Complete the sentences using the words from the list. Some of the words may be used more than once each.

Chimps are _____ closely related to gorillas than orangutans.
The common ancestor of chimps and gorillas is _____ recent than the common ancestor of gorillas and orangutans.

Words: less, more.

....

Q40: What process relies on the use of computer and statistical analyses to compare genome sequence data?

....

Q41: Comparison of genomes reveals that many genes are highly _____ across different organisms.

....

Q42: Current research is investigating how people's genetics affects their responses to drugs. What is this field of medicine known as?

Problem solving

An investigation was carried out to determine the effect of gamma radiation on germination in chickpeas.

Two varieties of chickpeas were exposed to gamma irradiation doses of 0 to 900 Gy (at 100 Gy intervals). Thirty seeds from each group were placed in a petri dish with moist filter paper, and left in an incubator at 25°C for 7 days. The experiment was repeated three times.

The average results from all three experiments are shown below.

Gamma radiation dose (Gy)	Percentage germination (%)	
	Kabuli chickpea	Desi chickpea
0	99	98
100	98	99
200	91	99
300	90	99
400	89	97
500	72	97
600	70	99
700	68	96
800	39	98
900	53	97

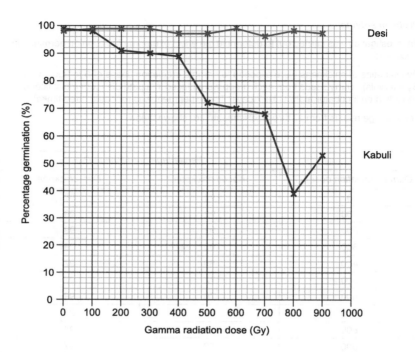

Q43: What term describes the group of seeds which were not exposed to gamma radiation (0 Gy)?

....

Q44: Why were the group of seeds not exposed to gamma radiation included in the study?

....

Q45: Name one factor, not already mentioned, which must be kept constant when setting up this experiment.

....

Q46: Describe the effect of increasing gamma radiation dose on percentage germination in kabuli chickpeas.

....

Q47: Predict the percentage germination of kabuli chickpeas if they were exposed to gamma radiation of 1000 Gy.

....

Q48: Draw one conclusion from the results of the experiment.

Unit 2: Metabolism and Survival

1	Metabolic pathways	133
	1.1 Introduction to metabolic pathways	135
	1.2 Membranes and metabolic pathways	135
	1.3 Metabolic pathways	136
	1.4 Enzyme action	137
	1.5 Control of enzyme activity	142
	1.6 Learning points	146
	1.7 Extended response question	148
	1.8 End of topic test	148
2	Cellular respiration	153
	2.1 The role of ATP	155
	2.2 The chemistry of respiration	157
	2.3 Learning points	163
	2.4 Extended response question	164
	2.5 End of topic test	164
3	Metabolic rate	167
	3.1 Measuring metabolic rate	168
	3.2 Oxygen delivery	169
	3.3 Learning points	171
	3.4 End of topic test	172
4	Metabolism in conformers and regulators	173
	4.1 Introduction	174
	4.2 Conformers	174
	4.3 Regulators	177
	4.4 Negative feedback control of body temperature	179
	4.5 Learning points	185
	4.6 Extended response question	186
	4.7 Extension materials	186
	4.8 End of topic test	188

5 Maintaining metabolism ... 189
- 5.1 Introduction ... 190
- 5.2 Dormancy ... 190
- 5.3 Migration ... 193
- 5.4 Learning points ... 196
- 5.5 End of topic test ... 196

6 Environmental control of metabolism ... 199
- 6.1 Microorganisms ... 200
- 6.2 Growth of microorganisms ... 204
- 6.3 Patterns of growth ... 207
- 6.4 Penicillin production ... 211
- 6.5 Learning points ... 212
- 6.6 Extended response question ... 213
- 6.7 Extension materials ... 213
- 6.8 End of topic test ... 215

7 Genetic control of metabolism ... 219
- 7.1 Improving wild strains of microorganisms ... 221
- 7.2 Mutagenesis ... 221
- 7.3 Selective breeding ... 222
- 7.4 Recombinant DNA ... 222
- 7.5 Recombinant DNA technology ... 223
- 7.6 Bovine somatotrophin (BST) ... 226
- 7.7 Learning points ... 228
- 7.8 Extension materials ... 229
- 7.9 End of topic test ... 230

8 End of unit test ... 231

Unit 2 Topic 1

Metabolic pathways

Contents

1.1 Introduction to metabolic pathways 135
1.2 Membranes and metabolic pathways 135
1.3 Metabolic pathways 136
1.4 Enzyme action 137
    1.4.1 Enzyme properties 137
    1.4.2 Enzymes and activation energy 140
1.5 Control of enzyme activity 142
    1.5.1 Controlling enzyme activity 142
    1.5.2 Competitive inhibition 142
    1.5.3 Non-competitive inhibition 143
    1.5.4 Feedback inhibition 144
    1.5.5 Experimental evidence of altering enzyme activity 146
1.6 Learning points 146
1.7 Extended response question 148
1.8 End of topic test 148

Prerequisites

You should already know that:

- the cell membrane consists of phospholipids and proteins and is selectively permeable;
- enzymes speed up cellular reactions and are unchanged in the process;
- enzymes function as biological catalysts and are made by all living cells;
- they speed up cellular reactions and are unchanged in the process;
- the shape of the active site of enzyme molecules is complementary to a specific substrate.

Learning objective

By the end of this topic, you should be able to:

- understand that metabolism is the sum total of all chemical reactions taking place within an organism;
- describe how reactions occur in sequences, each one mediated by a specific enzyme;
- describe these sequences as pathways and explain how they are linked and may be reversed;
- explain the differences between anabolic and catabolic pathways;
- describe the functions of protein embedded in the plasma membrane;
- understand that metabolic pathways are controlled by enzymes within the pathway;
- state that the rate of a pathway's reaction is controlled by the rates of the enzymes' reaction;
- describe the relationship between the shape (configuration) and activity of an enzyme;
- describe the 'induced fit' model of enzyme action;
- explain the role and function of the active site;
- explain what is meant by the term 'activation energy' and state that enzymes lower the activation energy of a reaction;
- explain how the concentrations of both substrate and end product affect the rate of enzyme action;
- explain why some genes are constantly expressed and how they are regulated;
- describe inhibition of enzymes in terms of a change in configuration;
- explain how metabolic pathways can be controlled through competitive and non-competitive inhibition of enzymes;
- explain how metabolic pathways can be controlled by feedback inhibition of enzymes.

1.1 Introduction to metabolic pathways

Metabolism is the term used to describe the enormous number of integrated and complex biochemical reactions that occur in an organism. These reactions are ordered into pathways and controlled at each stage by an enzyme. By means of these metabolic pathways, the cell is able to transform energy, degrade macromolecules and synthesise new organic molecules that are needed for life.

A **catabolic** reaction releases energy through the breakdown of a large molecule into smaller units (cellular respiration is a good example of this). An **anabolic** reaction uses energy to build small molecules into large ones, such as the synthesis of a protein from amino acids.

Many of the pathways are reversible, but some are not. For those that have stages that cannot be reversed, there are often alternative pathways that can overcome the blockage.

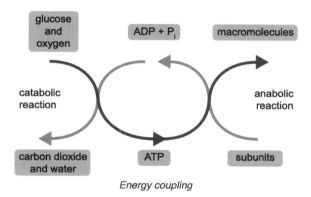

Energy coupling

1.2 Membranes and metabolic pathways

Membranes are vital to the activity of all cells. The plasma membrane separates the cell from its immediate environment whilst other membranes, similar in structure to the plasma membrane, divide the contents of the eukaryotic cell into specialised compartments. Molecules on the surface of the plasma membrane are involved in cell communication, and the membrane system as a whole is essential for transport both within cells and between cells.

The plasma membrane surrounding the cell is approximately 8nm wide, it is **selectively permeable**, and its unique structure determines both its function and physical characteristics.

Currently, the accepted model of membrane structure is the fluid mosaic model. This states that the membrane is made up of a bilayer of phospholipids with proteins embedded in it. The fluid mosaic model, illustrated below, highlights the complexity of the membrane, which is dynamic in nature.

136 UNIT 2. METABOLISM AND SURVIVAL

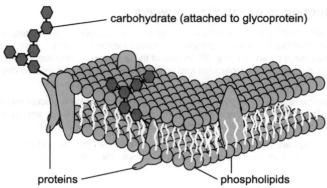

Fluid mosaic model of membrane structure

The membrane has proteins dispersed and embedded in the phospholipid bilayer that vary in both structure and function. The variety of functions carried out by membrane proteins are described below.

- *Channel (pore) proteins* - these proteins allow specific molecules and ions to pass through the membrane, for example, a protein channel found in the plasma membrane allows chloride ions (Cl^-) to cross the membrane.

- *Carrier (pump) proteins* - as the name suggests, carrier proteins bind to specific molecules or ions temporarily, enabling them to cross the membrane. This involves a change to the conformation of the carrier protein, which may require energy provided by ATP. The sodium-potassium pump is an example of a carrier protein involved in the transport of ions.

- *Enzymes* - some proteins in the membrane catalyse a specific reaction. Some receptor proteins have enzymatic activity, in which the cytoplasmic portion of the protein catalyses a reaction in response to binding by a ligand.

- *Structural support* - some membrane proteins are linked to the cytoskeleton and help to maintain the shape of the cell.

1.3 Metabolic pathways

Metabolism is the term used to describe all of the chemical reactions that occur within an organism. A metabolic pathway is a sequence of reactions that is controlled by enzymes that change one **metabolite** to another.

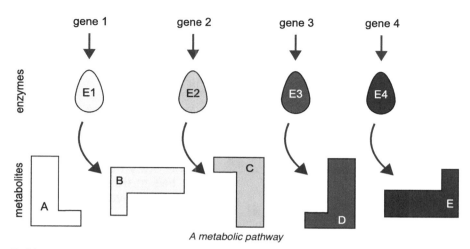

A metabolic pathway

Problems occur in metabolic pathways if the enzymes are not synthesised correctly, due to mutations in the genes that code for them. The next reaction in the pathway is then unable to occur and the intermediate metabolite builds up in the system.

Metabolic pathways are controlled by altering the presence and/or activity of key enzymes within the pathway. The regulation is brought about by signalling molecules from within the cell or from other cells.

1.4 Enzyme action

This section covers enzyme properties, enzymes and activation energy, and enzymes groups and multi-enzyme complexes.

1.4.1 Enzyme properties

Enzymes are three-dimensional **biological catalysts** comprising of globular protein molecules that are only produced in living organisms. They possess a small region called the **active site** where the substrate binds and the reaction occurs, and are specific in the reactions that they catalyse (one enzyme, one substrate).

Enzyme activity conforms to the **induced fit** model. The substrate molecule induces a slight change in the shape of the active site to allow the substrate molecule to fit perfectly. The change in shape of the active site facilitates the reaction by correctly orienting the reactants. After the reaction is complete, the products have a low affinity for the active site and are released; the active site resumes its normal shape and the enzyme is free to attach to more substrate molecules. This can be summarised as the catalytic cycle. A space-filling model of the enzyme hexokinase is shown below. This enzyme uses ATP to phosphorylate glucose to glucose-6-phosphate. The active site lies in a groove (as labelled), which closes when glucose is bound to the enzyme. This facilitates the reaction and increases the efficiency with which the enzyme binds ATP.

Space-filling model of the enzyme hexokinase showing the location of the active site

Enzymes are proteins, which means that their activity and/or structure will be affected by changes in conditions such as temperature and pH. The rate of an enzyme-catalysed reaction is also affected by:

- the concentration of the enzyme;
- the concentration of the substrate.

Since enzymes are catalysts, they:

- are required only in relatively small amounts;
- remain unchanged at the end of a reaction.

Hexokinase and the induced fit model of enzyme activity: Interactive 3D model   Go online

An activity that shows a molecular model of hexokinase (dimer) as an interactive 3D model is available in the online materials at this point. The following illustration gives an idea of what to expect. The position of a glucose molecule in the upper enzyme molecule is shown.

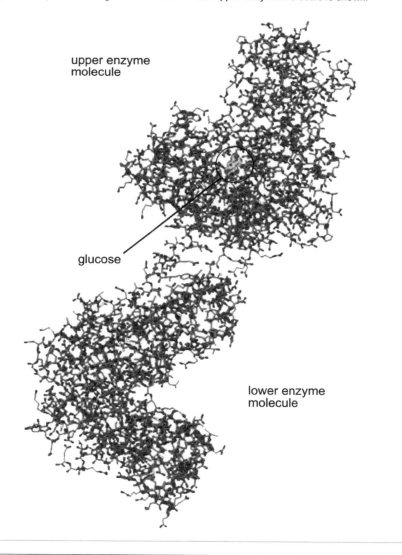

140 UNIT 2. METABOLISM AND SURVIVAL

Enzyme properties: Question Go online

The following illustrates the induced fit model of enzyme activity as applied to hexokinase.

Q1: Complete the diagram of the catalytic cycle for hexokinase by inserting the labels in the appropriate places.

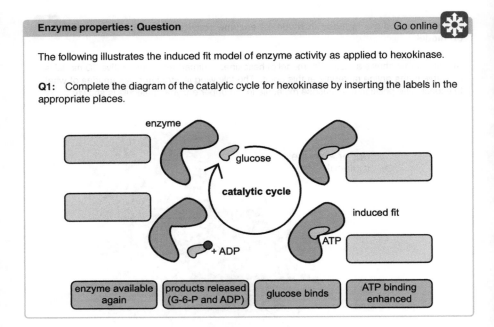

1.4.2 Enzymes and activation energy

Chemical reactions can involve the build up or a break down of a substance. In either case, the energy required to initiate the reaction is known as the **activation energy**.

If a catalyst is absent, the energy required to cause a chemical reaction is quite large and the speed of the reaction extremely slow. The presence of a catalyst ensures that the energy requirement is lowered and that the reaction takes place faster. In living systems, an enzyme lowers the activation energy by forming an enzyme-substrate complex that accelerates the rate of reaction. It is like rolling a boulder down a hill, but having to push it up a small hump first - this initial push takes energy, but after that the boulder rolls on. Enzymes make the small hump even smaller!

TOPIC 1. METABOLIC PATHWAYS

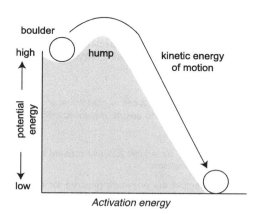

Activation energy

The effect of an enzyme on the activation energy of a reaction is shown below.

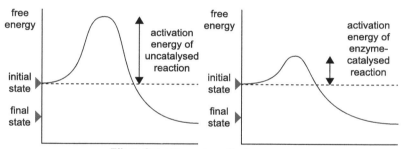

Effect of an enzyme on activation energy

Enzymes work by:

- bringing the substrates of a reaction close together (at the active site) so that they can react;
- lowering the activation energy of the reaction, so reactions can occur.

Enzyme action: Visualisation   Go online

A simple enzyme-substrate reaction is modelled below.

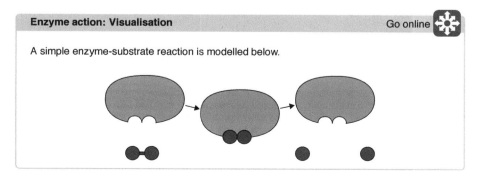

1.5 Control of enzyme activity

This section covers controlling enzyme activity, competitive inhibition, non-competitive inhibition, feedback inhibition and experimental evidence of altering enzyme activity.

1.5.1 Controlling enzyme activity

The number of enzyme-catalysed reactions in a cell is enormous, so the presence and number of enzyme molecules must be tightly controlled to ensure metabolic efficiency. Regulation of enzyme activity can be achieved in several ways:

- control of the number of enzyme molecules actually present in the cell - this is generally achieved at the level of gene expression;
- compartmentalisation, for example the enzymes required for the citric acid cycle and the electron transfer chain (two stages of respiration you will learn about in the next section) are contained in the mitochondria;
- change of enzyme shape - by far the most effective way of regulating an enzyme is to change its shape and therefore enzyme efficiency: a change in shape may either reduce or enhance enzyme activity, depending on the precise events taking place.

1.5.2 Competitive inhibition

Competitive inhibitors bind at the active site preventing the substrate from binding. This type of inhibition can be reversed by increasing the concentration of the correct substrate in the reaction.

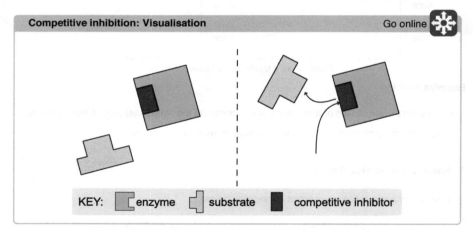

A good example of a reaction where the enzyme is subject to competitive inhibition is the conversion of succinate to fumarate in the citric acid cycle. This reaction is catalysed by the enzyme succinate dehydrogenase. However, if both succinate and malonate (which are very similar in structure) are present in the reaction vessel, they will compete for the active site of the enzyme. This reduces the rate of reaction because some of the active sites of the enzyme molecules are being occupied by the malonate. To increase the rate of reaction again, more succinate is added to the reaction vessel, ensuring that the enzyme is more likely to collide with the correct substrate molecule.

TOPIC 1. METABOLIC PATHWAYS

Competitive inhibition: Questions Go online

Q2: Explain the importance of the active site.

....

Q3: How is a competitive inhibitor related to the substrate of an enzyme-catalysed reaction?

....

Q4: How can competitive inhibition be overcome in experimental situations?

1.5.3 Non-competitive inhibition

Non-competitive inhibitors bind away from the active site but change the shape of the active site preventing the substrate from binding. Non-competitive inhibition cannot be reversed by increasing substrate concentration.

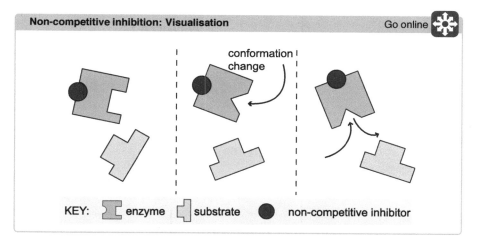

Non-competitive inhibition: Visualisation Go online

KEY: enzyme substrate non-competitive inhibitor

1.5.4 Feedback inhibition

Feedback inhibition occurs when the end-product in the metabolic pathway reaches a critical concentration. The end-product then inhibits an earlier enzyme, blocking the pathway, and so prevents further synthesis of the end-product.

An example takes place during the production of the amino acid isoleucine in bacteria, where the initial substrate is threonine which is converted by five intermediate steps to isoleucine. As isoleucine begins to accumulate, it binds to and inhibits the first enzyme in the pathway, thereby slowing down its own production. By this mechanism of end-product, or negative feedback, inhibition, the cell does not produce any more isoleucine than is necessary.

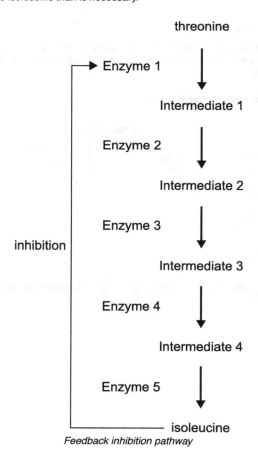

Feedback inhibition pathway

Feedback inhibition: Visualisation

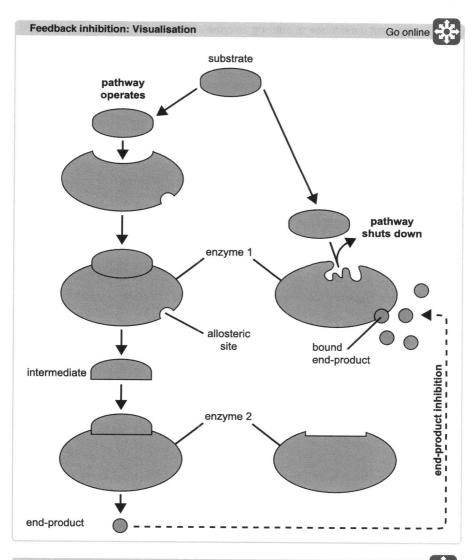

Feedback inhibition: Question

Q5: What is the advantage of end-product inhibition to the cell?

1.5.5 Experimental evidence of altering enzyme activity

The table below shows the results obtained in an experiment to investigate the effect of enzyme concentration on the rate of a reaction. The reaction that was investigated is the breakdown of starch into the disaccharide maltose by α-amylase. Starch hydrolysis was measured by the change in iodine staining (as determined using a spectrophotometer). The solutions of starch, phosphate buffer, and amylase were added to seven labelled test tubes.

Tube	Starch (0.02%, w/v) (ml)	Phosphate buffer (ml)	Amount of amylase added (ml)	μg amylase in reaction	Reaction rate (μg starch min^{-1})
1	5.0	4.5	0.5	50	34.7
2	5.0	4.0	1.0	100	50.0
3	5.0	3.5	1.5	150	56.5
4	5.0	3.0	2.0	200	73.9
5	5.0	2.5	2.5	250	82.1
6	5.0	5.0	0.0	0	0
7	0.0	5.0	5.0	500	0

The effect of amylase concentration on the rate of hydrolysis of starch to maltose

Q6: Draw a graph of the reaction rate against μg amylase in the reaction.

Q7: What is the effect of an increase in the amount of enzyme on the rate of the reaction.

Q8: What does the graph that you have drawn indicate about the reaction?

Q9: Explain the purpose of tubes 6 and 7.

Q10: What will happen to the reaction rate as the amount of amylase is increased above 250μg?

1.6 Learning points

Summary

- Metabolism is the term given to all the reactions that take place in the cell.
- Reactions in cells are controlled and co-ordinated by enzymes.

TOPIC 1. METABOLIC PATHWAYS

Summary continued

- Enzyme reactions do not take place in isolation but in pathways.
- Many of these pathways are reversible, but some are not.
- Where pathways are irreversible, or energetically unfavourable, alternative pathways are usually available.
- Reactions within metabolic pathways can be anabolic or catabolic. Anabolic reactions build up large molecules from small molecules and require energy. Catabolic reactions break down large molecules into smaller molecules and release energy.
- Most of these reactions are linked; the energy given off by one pathway is used to power another.
- Proteins within the membranes perform several functions, such as pores, embedded enzymes and channels.
- Metabolic pathways are controlled by a series of enzyme reactions.
- The rate of metabolism is dictated by the rate at which the enzymes work.
- Enzyme activity is closely related to its shape.
- Because enzymes are made from protein, their structure is flexible.
- When an enzyme and its substrate come together, the shape of the active site of the enzyme changes to allow a tighter fit with the substrate. This is called the 'induced fit' model.
- The active site of an enzyme creates an energetically favourable environment for the reaction to take place and lowers the activation energy.
- When the products are produced, they leave with a low affinity for the active site.
- The rate of enzyme reaction is affected by the concentration of the substrate and the end product.
- Most metabolic pathways are reversible. The direction will often depend on the quantity of substrate or the end product.
- Some genes, such as those for metabolism, are expressed continuously.
- These enzymes are always present, and are controlled through their rates of reaction.
- Enzymes can be inhibited by the binding of other particles.
- Non-competitive inhibitors bind away from the active site but change the shape of the active site, preventing the substrate from binding. Non-competitive inhibition cannot be reversed by increasing substrate concentration.
- Competitive inhibitors bind at the active site, preventing the substrate from binding. Competitive inhibition can be reversed by increasing substrate concentration.
- Feedback inhibition occurs when the end-product in the metabolic pathway reaches a critical concentration. The end-product then inhibits an earlier enzyme, blocking the pathway, and so prevents further synthesis of the end-product.

© HERIOT-WATT UNIVERSITY

1.7 Extended response question

The activity which follows presents an extended response question similar to the style that you will encounter in the examination.

You should have a good understanding of the control of the enzyme activity by inhibition before attempting the question.

You should give your completed answer to your teacher or tutor for marking, or try to mark it yourself using the suggested marking scheme.

Extended response question: The control of the enzyme activity by inhibition

Give an account of the control of the enzyme activity by inhibition. *(8 marks)*

1.8 End of topic test

End of Topic 1 test — Go online

Q11: The following diagram shows a metabolic pathway controlled by enzymes. The genes which code for each enzyme in the pathway are also shown.

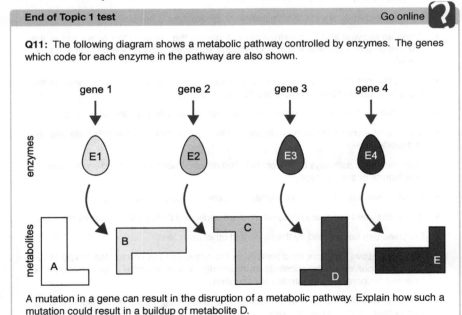

A mutation in a gene can result in the disruption of a metabolic pathway. Explain how such a mutation could result in a buildup of metabolite D.

....

Q12: What is metabolism?

....

TOPIC 1. METABOLIC PATHWAYS

Q13: Reactions which release energy are said to be _____ and reactions which require energy are described as _____. *(Choose between 'anabolic' and 'catabolic' for each gap.)*

Q14: Name the substances which control metabolic pathways.

Q15: Describe the role of pore proteins in the plasma membrane.

Q16: The following diagram shows the molecules involved in an enzyme controlled reaction.

Which list of molecule names identifies labels A, B and C from the diagram in order?

a) Non-competitive inhibitor, enzyme, competitive inhibitor.
b) Non-competitive inhibitor, substrate, competitive inhibitor.
c) Competitive inhibitor, substrate, non-competitive inhibitor.
d) Competitive inhibitor, enzyme, non-competitive inhibitor.

Q17: With respect to enzyme activity, what is meant by the term 'induced fit'?

a) An inhibitor binds to the active site preventing the substrate from binding.
b) Hydrogen and ionic bonds hold the substrate in the active site.
c) The correct conditions are created for substrate and enzyme to interact.
d) When the substrate binds to the active site, the shape of the active site is changed.

Q18: Enzymes _____ *(choose between 'increase' and 'lower')* the activation energy and release products with a _____ *(choose between 'high' and 'low')* affinity for the active site.

Q19: The following diagram shows the process of feedback inhibition.

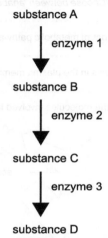

If substance D were to bring about feedback inhibition, which of the following will it interact with?

a) Enzyme 1
b) Enzyme 3
c) Substance A
d) Substance C

TOPIC 1. METABOLIC PATHWAYS

Q20: The figure below shows the progress of an enzyme-catalysed reaction.

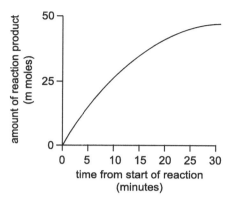

When is the rate of the reaction at its highest?

a) 0 to 5 minutes from the start of the reaction.
b) 5 to 15 minutes from the start of the reaction.
c) 15 to 25 minutes from the start of the reaction.
d) 25 to 30 minutes from the start of the reaction.

....

Q21: In an enzyme-catalysed reaction, the amount of product decreased when substance X was added. The addition of more substrate did not increase the amount of product. Substance X could be:

a) product
b) a competitive inhibitor
c) a non-competitive inhibitor
d) a cofactor

....

Q22: _____ inhibitors decrease the activity of an enzyme by binding to the active site.

....

Q23: The following illustrates a metabolic pathway.

$$A \rightarrow B \rightarrow C \rightarrow D \rightarrow E \rightarrow F$$

Assuming that molecule F regulates the first reaction in the pathway (the formation of B from A), which of the following would describe its mode of action?

a) Coenzyme activation
b) Feedback inhibition
c) Competitive inhibition
d) Non-competitive inhibition

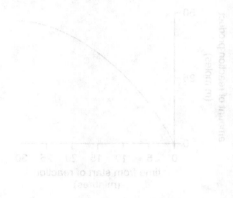

Unit 2 Topic 2

Cellular respiration

Contents

2.1 The role of ATP 155
2.2 The chemistry of respiration 157
    2.2.1 Glycolysis 157
    2.2.2 Citric acid cycle 158
    2.2.3 The electron transport chain 160
    2.2.4 Fermentation 161
    2.2.5 Measuring the rate of respiration 162
2.3 Learning points 163
2.4 Extended response question 164
2.5 End of topic test 164

Prerequisites

You should already know that:

- the chemical energy stored in glucose must be released by all cells through a series of enzyme-controlled reactions called respiration;

- the energy released from the breakdown of glucose is used to generate ATP from ADP and phosphate;

- the chemical energy stored in ATP can be released by breaking it down to ADP and phosphate;

- ATP can be regenerated during respiration;

- each glucose molecule is broken down via pyruvate to carbon dioxide and water in the presence of oxygen;

- the breakdown of each glucose molecule via the fermentation pathway yields two molecules of ATP when oxygen is not present;

- in the absence of oxygen in animal cells, glucose is broken down into lactate via pyruvate;

- in the absence of oxygen in plant and yeast cells, glucose is broken down into alcohol/ethanol and carbon dioxide via pyruvate;

Prerequisites continued

- fermentation occurs in the cytoplasm;
- aerobic respiration starts in the cytoplasm and is completed in the mitochondria.

Learning objective

By the end of this topic, you should be able to:

- understand that cellular respiration is the core pathway in cells which delivers energy for cell metabolism;
- describe how glucose is broken down to ultimately deliver ATP;
- explain that ATP is used to transfer energy to carry out cell processes;
- explain the reversible nature of ATP production;
- describe how ATP is synthesised;
- describe glycolysis;
- describe the progression of respiration pathways, both in the presence and absence of oxygen;
- describe the citric acid cycle;
- understand that respiration is a series of enzyme mediated reactions;
- explain the importance of the products of the citric acid cycle;
- describe the electron transport chain as a membrane bound system;
- explain the role of dehydrogenase enzymes;
- understand the key role of NAD and their reduced forms;
- explain the role of oxygen in the electron transport chain and the consequences of its absence.

2.1 The role of ATP

ATP is essential to biological systems as it is the link between reactions that release energy (catabolic) and those that use energy (anabolic).

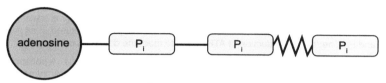

The structure of ATP

ATP is sometimes known as the 'energy currency' of the cell as it is spent during cellular work such as muscular contraction or the formation of proteins, and is 'banked' or stored when glucose is broken down during cellular respiration.

The illustration below shows the coupling of catabolic and anabolic reactions through ATP. You can see that energy to form ATP comes from the breakdown of food, and that the energy then contained in the ATP molecule is used to build up complex molecules from simple ones (such as proteins from amino acids).

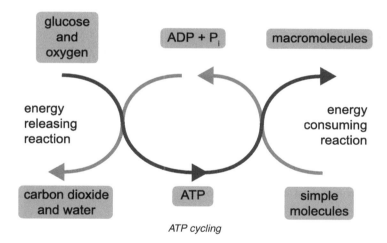

ATP cycling

UNIT 2. METABOLISM AND SURVIVAL

When an energy rich substance, such as glucose, is broken down in a living cell, it releases energy which is used to produce ATP. Many molecules of ATP are present in every living cell. Since ATP can rapidly be broken down to ADP + P_i (phosphate), it is able to make energy available for processes which need energy (e.g. muscular contraction, synthesis of proteins etc).

ATP is important, therefore, in that it provides the link between energy releasing reactions and energy consuming reactions. It provides the means by which chemical energy is transferred from one type of reaction to the other in a living cell. ATP is constantly manufactured in all living cells from ADP and P_i. The rate of production of ATP varies to meet the demands of the cell.

ATP also has a role in carrying out phosphorylation reactions within living cells. A phosphorylation reaction involves a phosphate group being added to a substrate. This is an enzyme controlled process. The formation of ATP from ADP and P_i is a phosphorylation reaction since a phosphate group is added to ADP to form ATP.

ATP can be used to phosphorylate other molecules within the cell. For example, during the first stage of respiration (glycolysis), ATP is broken down to ADP and P_i and the phosphate group is used to phosphorylate the substrate of glycolysis. This process initiates the reaction.

The role of ATP: Questions   Go online

Q1: Complete the diagram using the labels provided.

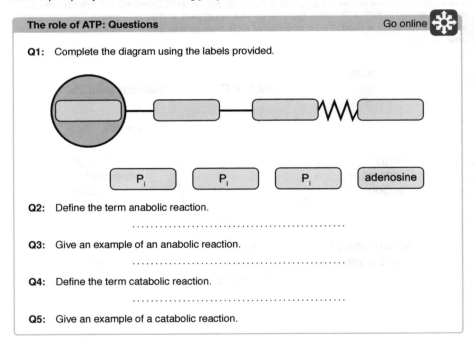

Q2: Define the term anabolic reaction.

....

Q3: Give an example of an anabolic reaction.

....

Q4: Define the term catabolic reaction.

....

Q5: Give an example of a catabolic reaction.

© HERIOT-WATT UNIVERSITY

2.2 The chemistry of respiration

Cellular respiration is a metabolic pathway. It consists of a series of enzyme-controlled reactions that release the energy contained in food, by oxidation.

There are three sets of reactions in cellular respiration:

1. **glycolysis**;
2. the **citric acid cycle**;
3. the **electron transport chain**.

2.2.1 Glycolysis

Glycolysis takes place in the cytoplasm of the cell and does not require oxygen. It is the breakdown, in a series of enzyme-catalysed reactions, of the sugar glucose into two molecules called pyruvate.

To start the process off, energy from two ATP molecules is needed. This can be thought of as an energy investment phase where ATP is used to phosphorylate intermediates in glycolysis. The series of reactions eventually produces four ATP molecules, so there is a net gain of two ATP from glycolysis (energy pay-off stage).

During the transformation of glucose into pyruvate, **dehydrogenase** enzymes remove hydrogen ions and electrons that are passed to a coenzyme called **NAD** which is reduced to form NADH. In a later process of cell respiration, the NADH will be used to produce ATP.

If oxygen is present pyruvate progresses to the **citric acid cycle**.

Glycolysis: Questions   Go online

Q6: Complete the diagram using the labels provided.

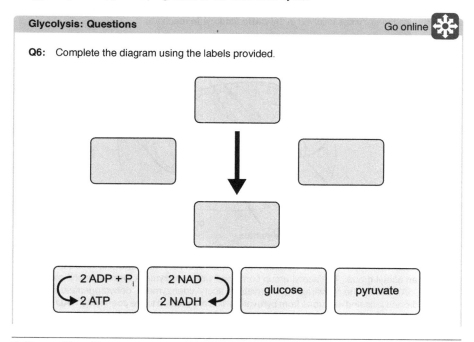

Q7: Where does glycolysis take place in the cell?

....

Q8: What is the net gain in ATP molecules from one glucose molecule during glycolysis?

a) 6
b) 2
c) 8
d) 4

....

Q9: Is oxygen required for glycolysis?

....

Q10: Name the final product of glycolysis.

2.2.2 Citric acid cycle

In the presence of oxygen, cell respiration continues on from **glycolysis** in the **mitochondria** of the cell.

Mitochondria (mitochondrion - singular) possess a double membrane. The inner membrane of each mitochondrion is folded into many cristae, which provide a large surface area. It is here that the reactions of the **electron transport chain** occur. The cristae project into a fluid-filled interior matrix which contains the enzymes involved in the **citric acid cycle** reactions.

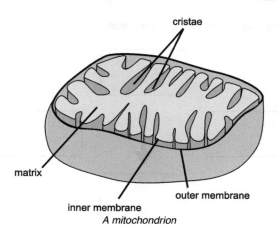

A mitochondrion

During the citric acid cycle, pyruvate diffuses into the matrix of the mitochondrion, where it is broken down into an acetyl group. The acetyl group combines with coenzyme A forming acetyl coenzyme A (acetyl CoA). During the conversion of pyruvate to acetyl coenzyme A, dehydrogenase enzymes remove hydrogen ions and electrons from pyruvate which are passed to the coenzyme **NAD**, forming NADH.

TOPIC 2. CELLULAR RESPIRATION

The acetyl group from acetyl coenzyme A combines with oxaloacetate to form citrate. During the citric acid cycle, citrate is converted through a series of enzyme-catalysed reactions back into oxaloacetate. In the process, both carbon (in the form of carbon dioxide) and hydrogen ions (along with electrons) are released.

Hydrogen ions and electrons become bound to NAD to form NADH. NADH will be used in the next stage of respiration to release energy for ATP production. Carbon dioxide diffuses out of the cell as a waste product and is expired from the organism by breathing out or by diffusion over the body surface.

Citric acid cycle: Questions   Go online

The following diagram illustrates the citric acid cycle.

Q11: Complete the diagram using the labels provided.

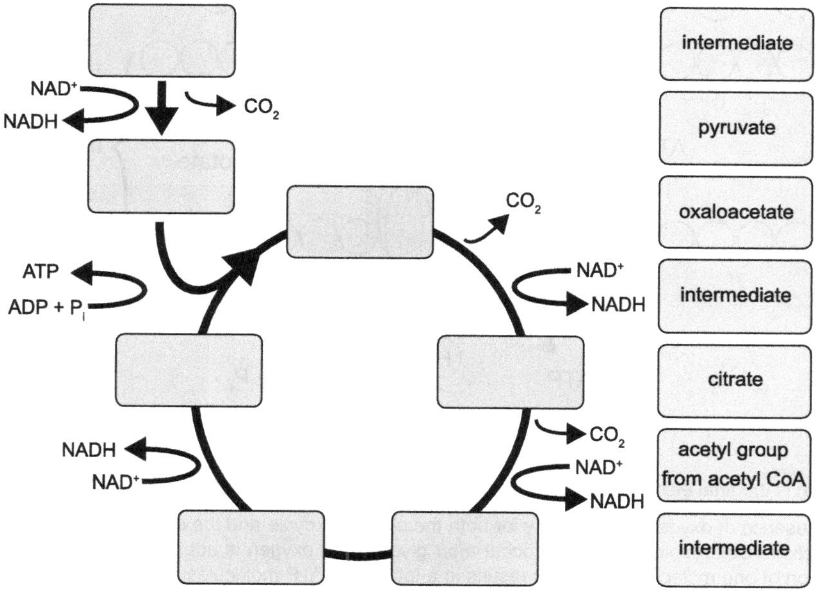

Labels provided:
- intermediate
- pyruvate
- oxaloacetate
- intermediate
- citrate
- acetyl group from acetyl CoA
- intermediate

Q12: Name the molecule produced when oxaloacetate combines with an acetyl group.

....

Q13: In which organelle does the citric acid cycle occur?

a) Chloroplast
b) Lysosome
c) Mitochondrion

Q14: Is citric acid cycle an aerobic or anaerobic process?

2.2.3 The electron transport chain

So far, many hydrogen ions and electrons have been transferred to **NAD** from **glycolysis** and the **citric acid cycle**. The electrons are passed to the **electron transport chain** on the inner **mitochondrial** membrane. The electrons are used to pump hydrogen ions across the mitochondrial membrane. The return flow of these hydrogen ions rotates part of the membrane protein **ATP synthase**. ATP synthase is an enzyme which catalyses the synthesis of ATP.

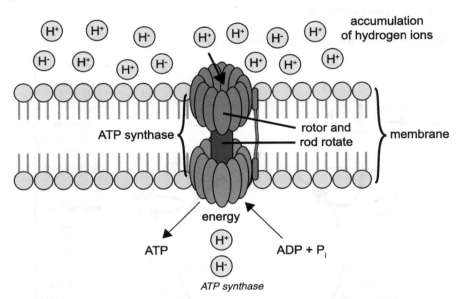

ATP synthase

Oxygen is the final electron acceptor, which combines with H ions and electrons, forming water.

The presence of oxygen is necessary for both the citric acid cycle and the electron transport chain to function. Cell respiration will not occur after glycolysis if oxygen is not present. The complete oxidation of one molecule of glucose results in a total of 38 ATP molecules. In glycolysis there is a net gain of two ATP molecules. In the citric acid cycle and during hydrogen ion transfer through the electron transport chain, 36 ATP molecules are produced, making a total of 38.

2.2.4 Fermentation

In the absence of oxygen, only **glycolysis** takes place and pyruvate follows a **fermentation** pathway in the cytoplasm. Fermentation results in much less ATP being produced than in aerobic respiration.

In animal cells, pyruvate is broken down into lactate. This can happen in the muscle cells of humans during vigorous exercise when all the oxygen available is used up. This process is reversible, as lactate can be converted back into pyruvate when oxygen becomes available again.

In plant and yeast cells, the fermentation pathway converts pyruvate into ethanol and carbon dioxide. This is an irreversible process because carbon dioxide is lost from the cell.

Anaerobic respiration: Questions Go online

The following provides a summary of anaerobic respiration in plants and animals.

Q15: Complete the diagram using the labels provided.

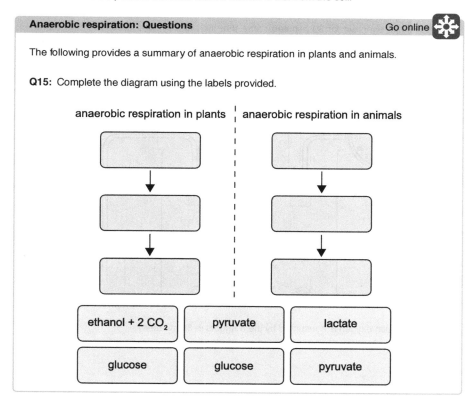

2.2.5 Measuring the rate of respiration

Measuring the rate of respiration — Go online

The **respirometer** on the right has been designed to measure the rate of respiration of some maggots. The respirometer on the left contains glass beads instead of maggots - this acts as a control.

As the maggots use oxygen for respiration, the level of the liquid in the glass tube will rise. Carbon dioxide produced by the maggots is absorbed by sodium hydroxide beads that have been placed underneath them.

The timer has been set for 30 minutes. After that time, the syringe above the maggots is pressed down to return the liquid in the tube to its initial level. The volume of oxygen used by the maggots can then be determined (the syringe is calibrated in units of 0.1 ml).

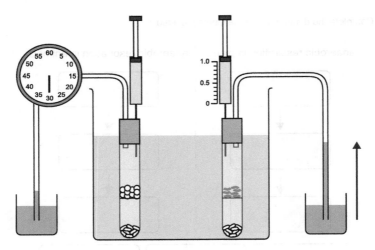

Q16: How much oxygen is consumed by the maggots in 30 minutes?

a) 0.1 ml
b) 0.2 ml
c) 0.3 ml
d) 0.4 ml

....

Q17: What is the rate of respiration (expressed as ml of oxygen consumed per hour)?

a) 0.3
b) 0.4
c) 0.5
d) 0.6

2.3 Learning points

Summary

- Glycolysis is the breakdown of glucose to pyruvate in the cytoplasm.

- ATP is required for the phosphorylation of glucose and intermediates during the energy investment phase of glycolysis. This leads to the generation of more ATP during the energy pay-off stage and results in a net gain of ATP.

- In aerobic conditions, pyruvate is broken down to an acetyl group that combines with coenzyme A forming acetyl coenzyme A.

- In the citric acid cycle, the acetyl group from acetyl coenzyme A combines with oxaloacetate to form citrate.

- During a series of enzyme controlled steps, citrate is gradually converted back into oxaloacetate which results in the generation of ATP and release of carbon dioxide.

- The citric acid cycle occurs in the matrix of the mitochondria.

- Dehydrogenase enzymes remove hydrogen ions and electrons and pass them to the coenzyme NAD, forming NADH. This occurs in both glycolysis and the citric acid cycle.

- The hydrogen ions and electrons from NADH are passed to the electron transport chain on the inner mitochondrial membrane.

- The electron transport chain is a series of carrier proteins attached to the inner mitochondrial membrane.

- During ATP synthesis electrons are passed along the electron transport chain releasing energy.

- This energy allows hydrogen ions to be pumped across the inner mitochondrial membrane. The flow of these ions back through the membrane protein ATP synthase results in the production of ATP.

- Finally, hydrogen ions and electrons combine with oxygen to form water.

- In the absence of oxygen, fermentation takes place in the cytoplasm.

- In animal cells, pyruvate is converted to lactate in a reversible reaction.

- In plants and yeast, ethanol and CO_2 are produced in an irreversible reaction.

- Fermentation results in much less ATP being produced than in aerobic respiration.

- ATP is used to transfer energy to cellular processes which require energy.

2.4 Extended response question

The activity which follows presents an extended response question similar to the style that you will encounter in the examination.

You should have a good understanding of the stages of cellular respiration before attempting the question.

You should give your completed answer to your teacher or tutor for marking, or try to mark it yourself using the suggested marking scheme.

Extended response question: The stages of cellular respiration

Give an account of the first two stages of cellular respiration under the following headings.

A) Glycolysis.

B) Citric acid cycle.

(8 marks)

2.5 End of topic test

End of Topic 3 test Go online

Q18: The production of ATP from ADP and P_i is called _____.

....

Q19: The break down of ATP releases _____, some of which is used in the synthesis of complex molecules.

....

Q20: Which statement referring to glycolysis is correct?

a) It is an anabolic reaction that requires oxygen.
b) It is a catabolic reaction that does not require oxygen.
c) It is an anabolic reaction that does not require oxygen.
d) It is a catabolic reaction that requires oxygen.

....

Q21: Which of the following does glycolysis produce?

a) Carbon dioxide
b) Citrate
c) Acetyl CoA
d) Pyruvate

TOPIC 2. CELLULAR RESPIRATION

Q22: Where in the cell does glycolysis take place?

a) In both the cytoplasm and the matrix of the mitochondrion.
b) On the surface of the inner mitochondrial membrane.
c) In the matrix of the mitochondrion.
d) In the cytoplasm.

Q23: The citric acid cycle occurs in the _____ of the mitochondrion.

Q24: During the citric acid cycle _____ combines with an acetyl group to form _____, this is gradually turned back into _____ by a series of _____ controlled reactions.

Q25: In the electron transport chain, what is the final acceptor of hydrogen?

a) Water
b) Carbon dioxide
c) Oxygen
d) NAD

Q26: Name the enzyme required to produce ATP

Q27: Name the coenzyme which transports hydrogen ions and electrons to the electron transport chain.

Unit 2 Topic 3

Metabolic rate

Contents

3.1 Measuring metabolic rate 168
3.2 Oxygen delivery 169
3.3 Learning points 171
3.4 End of topic test 172

Prerequisites

You should already know that:

- the pathway of blood through human heart, lungs and body;
- the structure of the human heart including the right and left atria and ventricles;
- red blood cells contain haemoglobin and are specialised to carry oxygen.

Learning objective

By the end of this topic, you should be able to:

- understand how the rate of metabolism can be measured;
- describe the mechanisms of delivery of oxygen in terms of the cardiovascular system of different animals.

3.1 Measuring metabolic rate

Metabolic rate is the quantity of energy used by the body over a given time. It is measured in kilojoules (or kilocalories). Metabolic rate can be measured in a number of ways, through the rate of oxygen consumption, carbon dioxide evolution or heat generation. The methods for detecting and measuring these factors range from very simple, low technology to more modern methods giving real time readings.

A common method used to calculate metabolic rate is to measure the quantity of oxygen used because in many cases respiration is aerobic. A figure frequently quoted is 350 litres per day for an average man. This equates to approximately 7000 kJ (or 1700 kcal) per day. Remember, the rate of oxygen uptake by an organism can be measured using a **respirometer**; for more information on respirometers look back to the interactivity in section 3.2.5 "Measuring the rate of respiration".

An organism's metabolic rate can also be calculated using a **calorimeter**. This piece of equipment monitors the heat generated by an organism and calculates the metabolic rate from the results collected. A simple set-up to measure heat generation by germinating peas is shown below. The boiled peas act as a control; there should be no temperature rise in this flask.

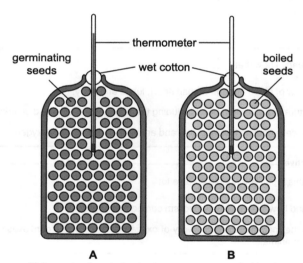

Using a simple calorimeter to measure generated heat

3.2 Oxygen delivery

Oxygen is consumed during aerobic respiration. If the rate at which energy is demanded rises, then so too must the rate of oxygen delivery. In simple organisms, oxygen can dissolve and diffuse through cell membranes. However, with increasing complexity, multi-cellular organisms must devise oxygen delivery systems. This gives rise to the cardiovascular systems seen in higher animals.

Organisms with high metabolic rates require more efficient delivery of oxygen to cells. Birds and mammals have higher metabolic rates than reptiles and amphibians, which in turn have higher metabolic rates than fish.

Fish have a single circulatory system consisting of one atrium and one ventricle. This model of a heart, which has an atrium to collect blood and a muscular ventricle to pump the blood around the body, is the basis for all vertebrates. In fish, the two-chambered heart results in a single circulatory system where blood passes through the heart once in each complete circuit of the body.

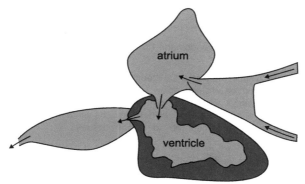

The heart of a fish

The hearts of amphibians and reptiles have evolved to a three-chambered organ, which has become necessary as there is a need to separate the deoxygenated blood from the oxygenated blood returning from the lungs. The circulatory system is described as an incomplete double circulatory system.

There are two atria in the heart which collect the blood. One collects blood from the body on the right (which is shown on the left in the diagram), and the other collects blood from the lungs on the left (which is shown on the right). Both chambers deliver blood to a single ventricle. There is relatively little mixing of the oxygenated and deoxygenated blood by means of a combination of timing of arterial contractions and the beginnings of a septum in the ventricle.

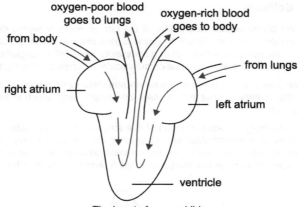

The heart of an amphibian

Birds and mammals have a high energy demand which requires efficient delivery of oxygen to the tissues.

The bird heart is essentially similar to that of the mammal. Blood returning from the body enters an atrium and is then pushed into the ventricle below. When the ventricle contracts, the blood leaves the heart and travels to the lungs where it is oxygenated. Blood leaves the lungs and travels back to the heart, entering the other atrium. Blood passes down into the ventricle below. When the ventricle contracts, the blood leaves the heart and travels to the body tissues.

Birds and mammals therefore have a complete double circulatory system consisting of two atria and two ventricles. Complete double circulatory systems enable higher metabolic rates to be maintained. There is no mixing of oxygenated and deoxygenated blood and the oxygenated blood can be pumped out at a higher pressure. This enables more efficient oxygen delivery to cells.

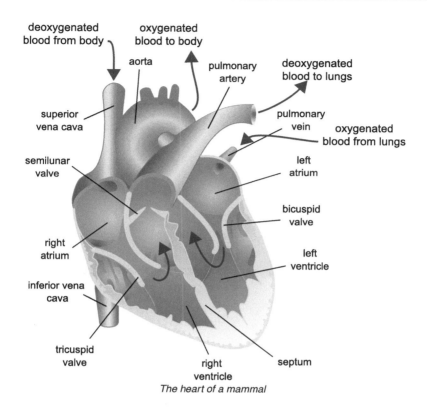
The heart of a mammal

3.3 Learning points

Summary

- The comparison of metabolic rates of organisms at rest is achieved by measuring either the oxygen uptake, carbon dioxide output or the heat produced.

- Metabolic rate can be measured using respirometers, oxygen probes, carbon dioxide probes and calorimeters.

- During aerobic respiration, oxygen needs to be delivered to respiring cells.

- Fish have a single circulatory system consisting of one atrium and one ventricle.

- Amphibians and most reptiles have an incomplete double circulatory system consisting of two atria and one ventricle.

> **Summary continued**
> - Birds and mammals have a complete double circulatory system consisting of two atria and two ventricles.
> - Complete double circulatory systems enable higher metabolic rates to be maintained. There is no mixing of oxygenated and deoxygenated blood and the oxygenated blood can be pumped out at a higher pressure. This enables more efficient oxygen delivery to cells.

3.4 End of topic test

End of Topic 4 test Go online

Q1: What instrument can be used to measure respiration?

Q2: Which of the following measurements do not allow metabolic rate to be calculated?

a) Heart rate
b) Heat generation
c) Oxygen consumption
d) Carbon dioxide production

Q3: The Mufflin formula for Resting Metabolic Rate (RMR) for a man is:

$RMR = 10w + 6.25h - 5a + 5$

Where: w = weight in kg; h = height in cm; a = age in years; and RMR is in kcal.

Calculate the RMR for a 45 year old man weighing 80 kg who is 170 cm tall.

Q4: A fish heart has ____ chambers.
An amphibian heart has ____ chambers.

Q5: What is the significance of the development of four-chambered heart?

Q6: A mammal has a _____ circulatory system whereas fish have a _____ circulatory system. *(choose from 'single' and 'double' for each gap)*

Unit 2 Topic 4

Metabolism in conformers and regulators

Contents

4.1 Introduction 174
4.2 Conformers 174
4.3 Regulators 177
4.4 Negative feedback control of body temperature 179
4.5 Learning points 185
4.6 Extended response question 186
4.7 Extension materials 186
4.8 End of topic test 188

Prerequisites

You should already know that:

- abiotic factors are non living factors such as temperature;
- the nervous system allows messages / signals to pass from one part of the body to another.

Learning objective

By the end of this topic, you should be able to:

- explain that an organism's ability to maintain its metabolic rate is affected by external abiotic factors;
- explain the differences between conformers and regulators, and the mechanisms they use in response to changes in the environment;
- explain the importance of regulating temperature during metabolism;
- describe the negative feedback control of temperature in mammals, including the role of the hypothalamus, nerves, effectors and skin.

4.1 Introduction

The metabolic rate of an organism will be affected by external conditions. How the organism responds to these changes in the external environment will alter according to the type of organism.

At the heart of metabolism are enzymes. It is the enzymes that regulate each and every reaction. Enzyme function is reliant on its three-dimensional configuration or shape. Anything that alters this will have an effect on the enzyme's ability to catalyse its specific reaction.

Of the many external environmental factors which may affect enzymes, pH, salinity and temperature are possibly the most obvious. For metabolism to remain efficient, enzymes should be in an environment that is maintained within fairly narrow parameters; extremes should be avoided as this would lead to denaturing of the enzymes.

Some organisms can do little to regulate their internal environment (conformers) while others can take extensive measures to regulate their internal environment (regulators).

4.2 Conformers

A conformer's internal environment is dependent upon its external environment. In other words, their internal conditions are controlled by environmental conditions. Conformers use behavioural responses to maintain optimum metabolic rate. Behavioural responses by conformers allow them to tolerate variation in their external environment to maintain optimum metabolic rate.

There are some advantages to being a conformer in that they have low metabolic costs, which means that little energy has to be used to drive mechanisms such as contractile vacuoles or other forms of active transport. There is, however, some disadvantage in that it will frequently restrict these organisms to narrow ecological niches and lower activity rates.

Two areas where this can be seen are responses to changes in osmolarity and temperature.

Osmolarity is a measure of the quantity of dissolved ions in water. Osmoconformers (frequently marine animals, such as squid) have body fluids that are at the same osmolarity as their surroundings. This, in turn, means that they have no need for structures such as kidneys, but it does leave them vulnerable to changes in their habitat.

TOPIC 4. METABOLISM IN CONFORMERS AND REGULATORS

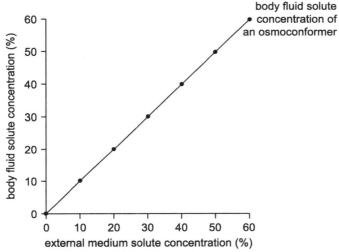

Osmolarity of body fluids and the environment in osmoconformers

If the conditions change rapidly, then cell or organism death will result either by water leaving or entering in an overwhelming manner.

Thermoconformers are described as animals that cannot regulate their body temperature internally. This includes most insects and reptiles. The most frequent method of control is that of adapted behaviour.

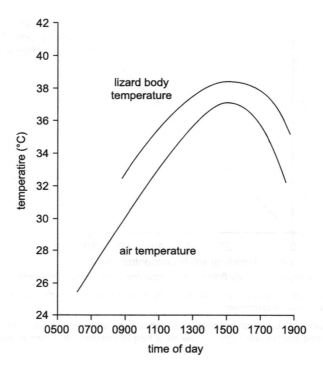

Body and external temperature in a thermoconformer, e.g. a lizard

TOPIC 4. METABOLISM IN CONFORMERS AND REGULATORS

The behaviours that are exhibited include:

- vaporisation (getting wet);
- convection (losing or gaining heat to an airflow);
- conduction (lying next to some colder or warmer surface or material);
- radiation (finding shade or lying on hot rocks).

The major advantage of this behaviour is that little energy is required to maintain a steady body temperature. As the food/energy requirement is much lower for thermoconformers, they can survive erratic food supplies and so increase their chances of survival. However, should the temperature rise, their need for food will also rise as the metabolic rate increases with the surrounding temperature.

4.3 Regulators

Regulators are organisms which are able to use metabolic means to regulate their internal environments in response to external environmental changes. In order to achieve this, large quantities of energy are required in order to power specialist organs and body systems. The advantage of this, however, is that it allows these organisms to live more independently of their external environment and occupy a wider range of ecological niches. The maintenance of an internal environment in a 'steady state' is called homeostasis. It requires energy to achieve homeostasis, therefore, regulators have higher metabolic costs than conformers.

An osmoregulator can maintain its osmolarity despite wide variations in its habitat. It can move between a wide variety of habitats and niches and not be adversely affected. This group includes arthropods and vertebrates.

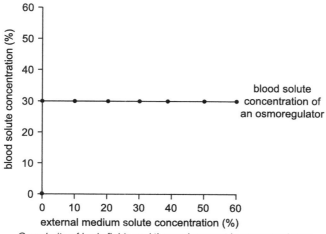

Osmolarity of body fluids and the environment in osmoregulators

178 UNIT 2. METABOLISM AND SURVIVAL

Fish have adapted to usually either marine or fresh water environments (although some can range between both) by using salt organs in the gills and kidneys. In marine habitats, the overwhelming problem is that there is a the tendency is for water to be drawn out of the body. In addition, there is a diffusion gradient by which salts will move into the fish. To overcome these problems, the fish drinks water constantly and excretes salt from its gills. It also produces a small volume of concentrated urine.

Freshwater fish have the opposite problem, which is that water is moving into the fish while it is losing salts to the external environment. In this instance, the mechanisms are reversed so that salt cells in the gills actively absorb salt from the environment and the kidneys produce large volumes of very dilute urine.

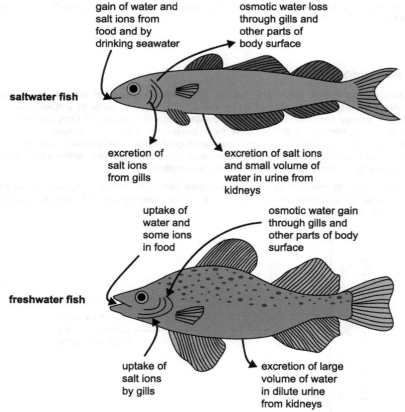

Control of blood salt concentration in saltwater and freshwater fish

Thermoregulators are organisms which can regulate their internal temperature, for example birds and mammals. For more information on thermoregulation read through the section.

4.4 Negative feedback control of body temperature

Feedback is used to regulate the response to a change in the environment. In **negative feedback**, when a condition changes, the opposite effect is produced by the body to return itself to normal. For example, if the internal temperature increases, the body responds by stimulating the processes to cool itself down.

The body is covered in **receptor** cells (both inside and outside) that detect changes in the environment. The change acts as a stimulus for the receptor cells, which then send a message to a control unit (usually the brain). A decision is made as to how the body is going to respond to the environmental change and a message is sent to the appropriate **effector** cells. These cells perform the response and return the body to its normal state. All of this happens very, very quickly so that almost as soon as a change occurs, a response is initiated.

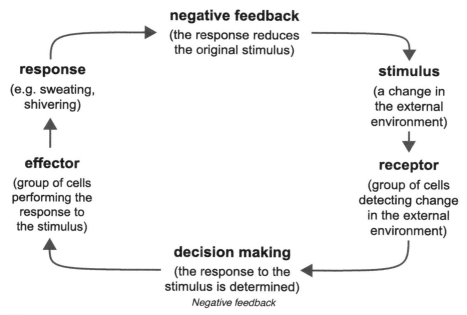

Negative feedback

The messages sent between the receptor and effector cells are hormonal messages or nerve impulses.

Regulating body temperature

The internal body temperature of a mammal needs to remain constant in order for the body to function properly. Although humans can withstand a wide range of environmental temperatures, we need to keep our internal body temperature as close to 37 °C as possible. Slight changes can cause major problems for the body and can even result in death. For example:

- a temperature increase denatures enzymes and blocks metabolic pathways;
- a temperature decrease slows metabolism and affects the functioning of the brain.

Different types of animals control their internal body temperature in different ways. Lizards, along with other reptiles, invertebrates, amphibians and fish, are thermoconformers. This means that they regulate their body temperature by absorbing heat from the surrounding environment. Therefore, their body temperature changes with that of the environment.

On the other hand, thermoregulators, such as mammals and birds, are able to maintain a constant body temperature regardless of the environmental temperature. They do not need to absorb heat from the surroundings because they gain heat energy from their own metabolism. Thermoregulators regulate their body temperature using homeostatic mechanisms and are able to live in a wider range of temperatures than thermoconformers (think about how widespread humans are in comparison with crocodiles).

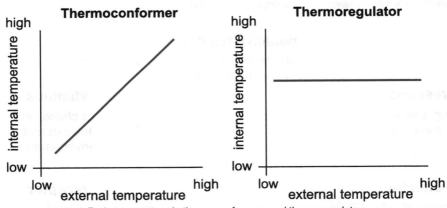

Body temperature in thermoconformers and thermoregulators

One of the roles of the **hypothalamus** is monitoring and responding to changes in body temperature - it is the body's thermostat, found in the brain. Thermoreceptors in the hypothalamus detect changes in the temperature of the blood, which correspond to temperature changes of the core of the body. The hypothalamus also responds to changes in the surface temperature of the body. These temperature changes are detected by millions of thermoreceptors in the skin which send messages, via nerve impulses, to the hypothalamus.

The hypothalamus processes the information it receives from the thermoreceptors and decides what response is needed (i.e. does it need to heat the body up or cool it down?). Electrical impulses are sent through nerves to the **effectors** which perform the chosen response to the stimulus. Feedback tells the body when its temperature has returned to normal.

TOPIC 4. METABOLISM IN CONFORMERS AND REGULATORS

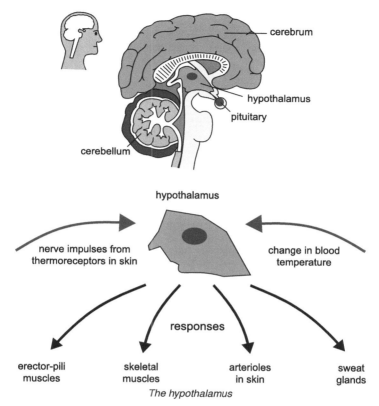

The hypothalamus

The skin is a highly complex organ that plays a very important role in regulating the temperature of a mammal. Not only does the skin contain **receptor** cells that detect temperature changes, but it also acts as an effector in response to nerve impulses from the hypothalamus.

When the hypothalamus detects an increase in body temperature it responds by taking measures to cool the body down (and vice versa). The main mechanisms employed by the skin to regulate the body temperature of a mammal are:

- **vasodilation** - the blood vessels (arterioles) that supply blood to the skin dilate (widen), increasing the amount of blood flowing to the skin. This increases the surface area from which heat can be lost to the environment by radiation.

- **vasoconstriction** - the arterioles that supply blood to the skin constrict (narrow), reducing the amount of blood flowing to the skin. As a result less heat is lost by radiation from the surface of the body.

Vasodilation and vasoconstriction

The following illustrates vasodilation and vasoconstriction.

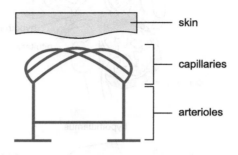

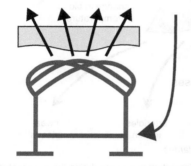

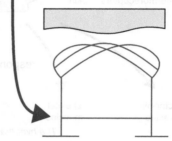

If a nerve impuse from the hypothalamus signals that the body temperature is hot, the arterioles dilate, sending more blood to the capillaries so that heat is lost from the surface of the skin.

If a nerve impuse from the hypothalamus signals that the body temperature is cold, the arterioles constrict, sending less blood to the capillaries so that the heat loss from the surface of the skin is minimised.

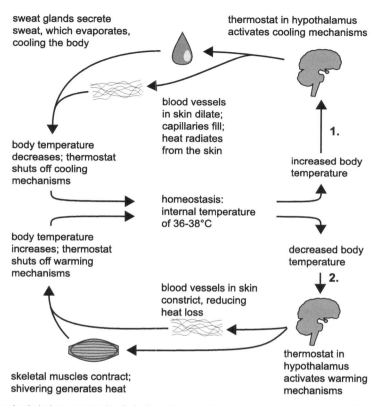

When the body is hot, sweat glands in the skin secrete sweat which cools the body down when it evaporates from the surface of the skin. When the body is cool, less sweat is produced to conserve heat.

Hairs are embedded in the sublayers of the skin and are attached to the skin epidermis by erector pili muscles. When the body is cold, nerve impulses from the hypothalamus contract the erector pili muscles, causing the hairs to stand up. This traps a layer of air close to the body, which acts as insulation. This mechanism is less effective in humans than it is in birds and furry mammals.

The body also employs other mechanisms to regulate its body temperature. For example, shivering warms the body up, as does increasing the rate of metabolism (this generates more heat energy). On the other hand, decreasing the rate of metabolism reduces the amount of heat energy produced by the body.

Not all temperature regulation mechanisms are physiological. When a person feels hot they will usually drink more water, take some of their clothes off or stay in the shade. However, when they feel cold they may have a hot drink, put more clothes on or hug a hot water bottle!

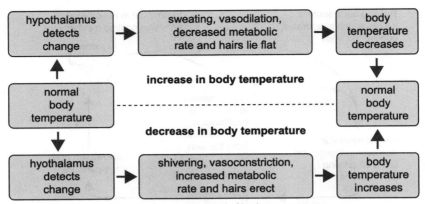

Negative feedback control of body temperature

Regulating body temperature: Question Go online

Q1: Complete the table with the temperature responses listed.

Decrease in body temperature	Increase in body temperature

Temperature responses: active sweat glands, decrease in metabolic rate, hair erector muscles contracted, hair erector muscles relaxed, inactive sweat glands, increase in metabolic rate, vasoconstriction, vasodilation.

Q2: Which part of the brain is responsible for regulating body temperature in mammals?

....

Q3: In which of the following ways does the body respond when its temperature falls?

a) Vasoconstriction and sweating
b) Contraction of hair erector muscles and vasodilation
c) Vasodilation and decreased rate of metabolism
d) Shivering and vasoconstriction

....

Q4: Why do you sweat and your skin become flushed during exercise?

4.5 Learning points

Summary

- The ability of an organism to maintain its metabolic rate is affected by the external environment.
- Abiotic factors in the external environment include temperature, pH and salinity.
- Conformers are organisms which do not regulate their internal environment by metabolic means.
- The internal environment of a conformer is dependent upon its external environment.
- There is an energetic advantage to this as they do not have to expend much metabolic energy.
- The result however is that they are confined to a narrow range of ecological niches.
- They can respond to a limited extent by behaviour responses to help maintain optimum internal conditions.
- Regulators maintain their internal environment regardless of external environment.
- This allows them to occupy a greater range of ecological niches.
- This level of control demands a high energy expenditure, with a considerable proportion of metabolic activity devoted to maintaining a steady state.
- The process is called homeostasis.
- To maintain a steady state negative feedback must be employed.
- Negative feedback occurs when a trend is reversed back to a 'normal level'. For example if body temperature rises above the set 'norm' then negative feedback returns it toward that 'norm'.
- For such a mechanism to work there have to be in place receptors to detect change in a given parameter (e.g. temperature), effectors to bring about any change, and a communication system between the two (usually nerves).
- It is important to regulate temperature (thermoregulation) for optimal enzyme activity and high diffusion rates to maintain metabolism.
- The hypothalamus is the temperature monitoring centre. Information is communicated by electrical impulses through nerves to the effectors, which bring about corrective responses to return temperature to normal.
- If body temperature increases above normal, processes such as sweating (body heat used to evaporate water in the sweat, cooling the skin), vasodilation (increased blood flow to the skin increases heat loss) and decreased metabolic rate (less heat produced) help to bring body temperature back down to normal.
- If body temperature decreases below normal, processes such as shivering (muscle contraction generates heat), vasoconstriction (decreased blood flow to skin decreases heat loss), hair erector muscles contract (traps layer of insulating air) and increased metabolic rate (more heat produced) help to bring body temperature back up to normal.

186   UNIT 2. METABOLISM AND SURVIVAL

4.6   Extended response question

The activity which follows presents an extended response question similar to the style that you will encounter in the examination.

You should have a good understanding of internal body temperature regulation in mammals before attempting the question.

You should give your completed answer to your teacher or tutor for marking, or try to mark it yourself using the suggested marking scheme.

Extended response question: Internal body temperature regulation in mammals

Give an account of how internal body temperature is regulated in mammals. *(7 marks)*

4.7   Extension materials

The material in this section is not examinable. It includes information which will widen your appreciation of this section of work.

Extension materials: ADH

The term osmoregulation describes the mechanisms by which the body maintains a constant level of water, ions and salts in its cells. Controlling the amount of water in the body is very important as it can be just as dangerous to be over-hydrated as dehydrated.

The kidneys act as osmoregulators in the human body. They respond to a hormone produced by the pituitary gland and reabsorb water back into the blood.

The concentration of water in the blood decreases if a person loses a lot of water, fails to take in enough water or consumes lots of salty food.

So, how does the body know that its cells need more water? There is a group of cells found in the hypothalamus, next to the pituitary gland in the brain. These act as the receptor cells that monitor the water concentration in the blood. If the concentration of water falls these receptor cells are stimulated and cause the pituitary gland to increase the secretion of a hormone into the bloodstream. When the kidneys detect this hormone they increase the rate at which they reabsorb water back into the bloodstream. As a result a smaller volume of more concentrated urine is produced by the body (to minimise water loss).

The hormone secreted by the pituitary gland is called antidiuretic hormone (ADH). (Its name describes its function: diuresis is the production of large volumes of dilute urine; ADH produces the opposite.) ADH works by making the ducts and tubules in the kidneys more permeable to water. Therefore, water is able to move (by osmosis) into the tissues and bloodstream from the kidneys more easily.

© HERIOT-WATT UNIVERSITY

To prevent you from becoming too dehydrated the receptors in the brain that increase the secretion of ADH also make you feel thirsty. Drinking, along with the re-absorption of water in the kidneys, helps the body to restore the normal blood water concentrations. However, if you drink too much liquid the body reduces the production of ADH and the kidneys re-absorb less water. This allows the body to remove the excess water - in other words, you have to go to the toilet a lot!

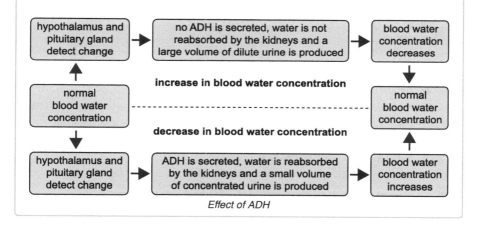

Effect of ADH

4.8 End of topic test

End of Topic 5 test   Go online

Q5: Which of the following factors do not affect the ability of an animal to maintain its metabolic rate?

a) Light intensity
b) pH
c) Salinity
d) Temperature

....

Q6: Explain why it is important to maintain an organism's body temperature within a relatively narrow range.

....

Q7: The internal environment of _____ is dependent upon the external environment. _____ control their internal environment.

(Choose between 'conformers' and 'regulators' for each gap.)

....

Q8: Give an advantage and a disadvantage of conformers over regulators.

....

Q9: Which part of the brain detects body temperature?

....

Q10: What term describes the process whereby the blood vessels immediately under the skin become narrower to restrict blood flow and reduce heat loss?

....

Q11: Which of the following occur when the body is too hot?

a) Decreased metabolic rate
b) Increased metabolic rate
c) Vasoconstriction
d) Vasodilation

....

Q12: How are signals sent from the temperature-monitoring centre of the brain to effectors?

Unit 2 Topic 5

Maintaining metabolism

Contents

5.1 Introduction 190
5.2 Dormancy 190
5.3 Migration 193
5.4 Learning points 196
5.5 End of topic test 196

> **Prerequisites**
>
> You should already know that:
>
> - an adaptation is an inherited characteristic that makes an organism well suited to survival in its environment/niche.

> **Learning objective**
>
> By the end of this topic, you should be able to:
>
> - describe the mechanisms and adaptations of organisms to survive in adverse conditions;
> - describe how reducing metabolic rate (dormancy) can be used as a strategy for surviving adverse conditions;
> - describe how relocation (migration) can be used as a method of avoiding adverse conditions.

5.1 Introduction

All organisms need to grow, develop and reproduce. Varying quantities of energy are devoted to each of these processes by organisms, and different strategies are employed to enhance the organism's chances of survival. Organisms also respond to changes in their external environment in different ways. In order to survive, organisms need to overcome the pressures of their environment and generate enough energy to complete their life-cycle.

When environmental conditions vary beyond the tolerable limits for normal metabolic activity, for example extremes of temperature or lack of water, organisms frequently resort to two types of survival mechanism: the first is to devise mechanisms to survive the condition (dormancy) and the second is to avoid them (migration).

5.2 Dormancy

One commonly employed method of ensuring survival involves **dormancy**. Dormancy is part of the lifecycle of some organisms to allow survival during a period when the costs of continued normal metabolic activity would be too high. The metabolic rate can be reduced during dormancy to save energy. During dormancy, there is also a decrease in heart rate, breathing rate and body temperature.

Dormancy can be predictive or consequential.

A predictive strategy allows dormancy to occur before the onset of unfavourable conditions. For example, decreasing temperature and day lengths are cues in seasonal environments that predict the onset of winter.

A consequential strategy enables the organism to react immediately to environmental cues. Organisms only enter a state of dormancy after they have been exposed to the adverse conditions. This is typically found in unpredictable environments where conditions may change very quickly. There is an enormous disadvantage to this, as a sudden change in conditions may result in high mortality rates. However, the organisms can delay dormancy until adverse conditions arise, meaning that they can make full use of the resources available in the habitat for as long as possible.

An organism may become dormant in response to changes in the environment or dormancy may be part of its life cycle. There are several different ways in which organisms save energy:

- **Daily torpor** involves the reduction of an organism's activity and metabolic rate for part of the day. Daily torpor often involves a reduction in heart rate and breathing rate. This allows organisms with high metabolic rates to save energy when they would not be able to find food. For example, house mice are active during the night and experience torpor through the day when it would be dangerous for them to be out in the open foraging for food.

- **Hibernation** is used by many organisms to escape cold weather conditions and scarce food supplies. The normal body functions of an organism change dramatically during hibernation. For example, the heart rate of the jumping mouse falls from 600 beats per minute to just 30 beats per minute. Animals prepare for hibernation by eating lots of food in the late summer and autumn. This builds up a layer of fat which keeps them warm and acts as a food source during the hibernation period. Hibernation can be either a predictive or consequential strategy. This form of dormancy is commonly seen in mammals such as hedgehogs, bears and dormice.

- **Aestivation** is a form of dormancy entered into by organisms in response to very hot and dry conditions. For example, the garden snail and some worms become dormant until moisture levels rise again. The snail retreats into its shell and seals the end and the worm coils up in a pocket of air surrounded by mucus. A more amazing example of aestivation is the lungfish, found in South America and Africa. This fish survives drought by burying itself in the mud on the river bed; the mud dries with the fish inside where it is able to survive until the next rainy season. Aestivation is a consequential strategy.

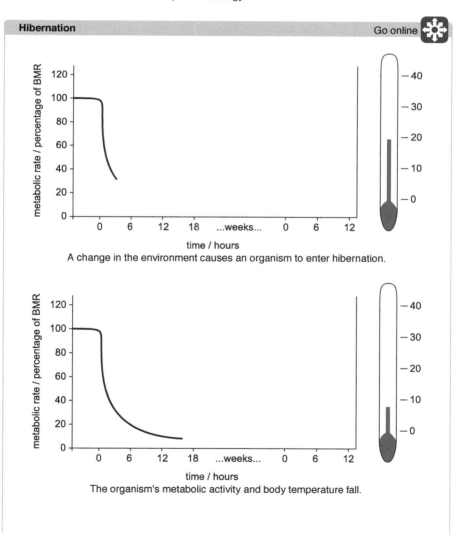

192 UNIT 2. METABOLISM AND SURVIVAL

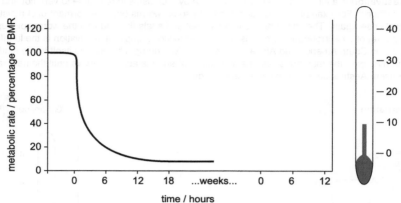

An organism may hibernate for several weeks or even months. During this period, its metabolic activity and body temperature remain constant, but at a lower level than in the active organism.

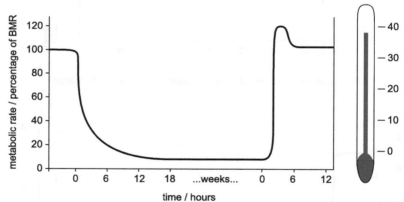

As it emerges from hibernation, the organism's metabolic activity and body temperature increase rapidly. Initially, it metabolises at a higher rate than normal to generate energy for its normal active functions.

© HERIOT-WATT UNIVERSITY

TOPIC 5. MAINTAINING METABOLISM

Dormancy: Questions   Go online

Q1: Complete the table using the descriptions listed.

Term	Definition
Hibernation	
Aestivation	
Daily torpor	

Description list:

- Dormancy in response to hot, dry conditions
- Dormancy in response to low temperatures
- A period of reduced activity in some animals with high metabolic rates

5.3 Migration

In addition to physiological responses, organisms use behavioural strategies to avoid adverse environmental conditions. Many organisms, such as swallows, whales and moose, **migrate** to areas with more favourable conditions. All migration involves a considerable investment of energy from the individuals concerned, but is beneficial in the long-term, as it allows them to relocate to a more suitable environment.

Observation of migration goes back at least to the times of the ancient Greeks. Although some of the explanations may seem a little strange today. Aristotle thought that robins turned into redstarts at the onset of summer and the barnacle goose grew as a fruit from trees and had an intermediate state as goose barnacles on drift wood.

As time progressed, more critical observation was able to demonstrate that a large variety of animals, mammals, birds, fish and insects made long journeys from typically feeding and breeding grounds to over-wintering areas and back again the following year. To improve accuracy, early tagging methods were employed in conjunction with capture and release techniques.

194 UNIT 2. METABOLISM AND SURVIVAL

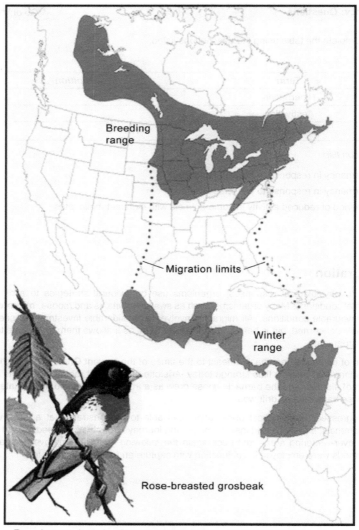

Rose-breasted grosbeak's summer and winter areas and migration routes

TOPIC 5. MAINTAINING METABOLISM

Capture and release technique

Advances in technology allow for radio tags which can be followed by VHF receivers or satellite. This allows real-time tracking of an individual. Such methods are now used with many species of animal.

Work carried out by Dr. Peter Berthold and others has now established that the ability to successfully migrate is most likely to be hereditary. By observing the behaviour of a species of bird which has populations that migrate and others that do not, it was found that after breeding between the two types a significant proportion of the offspring had the ability to successfully migrate. The fact that not all of the offspring received the ability suggests that there is more than one gene involved.

It was further found that this inbuilt ability was controlled by a circannual rhythm. That is these birds have an inbuilt 'clock' that measures time. It was thought to have a natural cycle of approximately ten months. It will most likely be coordinated by day length.

Migratory behaviour is thought to be influenced by both innate and learned behaviour. Innate behaviour is inherited from parents to offspring and is likely to be the biggest influence on successful migration. Learned behaviour is gained by experience. Learned behaviours may come from parents or other members of a social group.

5.4 Learning points

Summary

- Many environments experience a range of conditions which will not support life.
- To be able to survive such conditions, organisms have evolved strategies to offset the conditions.
- One strategy is to lower metabolic rate, which is known as dormancy.
- Dormancy is described as a period when growth, development and activity are temporarily suspended.
- Dormancy is either predictive (occurs before the onset of adverse conditions) or consequential (occurs after the onset of adverse conditions).
- Hibernation in mammals is mostly in response to the onset of winter, with reduced temperature and shorter day length.
- Aestivation, usually in summer, is induced by high temperature and lack of water.
- Some organisms which have small body size but high metabolic rates, e.g. hummingbirds, conserve energy by decreasing physiological activity over nights, a process known as daily torpor.
- Some organisms avoid adverse conditions by relocating to a more suitable (survivable) environment. This is migration.
- During migration, the energy that would have been used to survive adverse conditions is used to travel to another environment.
- Techniques have been developed to study long distance migration. These include tagging, radio tracking, capture and release, and direct observation.
- It is believed that the ability of animals to follow migratory pathways is a combination of inherited (innate) and learned behaviour.

5.5 End of topic test

End of Topic 6 test   Go online

Q2: Which strategy is used by an organism that enters a state of dormancy **after** it has been exposed to adverse conditions?

a) Complete dormancy
b) Consequential dormancy
c) Predetermined dormancy
d) Predictive dormancy

....

Q3: Dormancy entered into by organisms in response to very hot and dry conditions is called:

a) aestivation
b) daily torpor
c) hibernation

Q4: Dormancy during the winter, in which the body temperature and metabolic rate of the animal drops significantly, is called:

a) aestivation
b) daily torpor
c) hibernation

Q5: Aestivation is a form of:

a) consequential dormancy.
b) predictive dormancy.
c) both predictive and consequential dormancy.

Q6: Hibernation is a form of:

a) consequential dormancy.
b) predictive dormancy.
c) both predictive and consequential dormancy.

Q7: Dormice hibernate for five to seven months of the year. During this time, their heart rate can drop from 600 beats per minute to 30 beats per minute.
What is the percentage decrease in the heart rate of the dormouse during hibernation?

Q8: Successful migration depends on a combination of _____ and _____ behaviours.

Q9: Name one method of studying migration.

Unit 2 Topic 6

Environmental control of metabolism

Contents

6.1 Microorganisms 200
    6.1.1 Archaea 201
    6.1.2 Bacteria 201
    6.1.3 Fungi 202
    6.1.4 Protozoa 203
    6.1.5 Algae 204
6.2 Growth of microorganisms 204
6.3 Patterns of growth 207
6.4 Penicillin production 211
6.5 Learning points 212
6.6 Extended response question 213
6.7 Extension materials 213
6.8 End of topic test 215

Learning objective

By the end of this topic, you should be able to:

- describe the variety of microorganisms;
- explain their methods of metabolism and ecological range;
- learn how they can be exploited in industry;
- describe methods and conditions necessary to cultivate microbes;
- explain growth patterns.

6.1 Microorganisms

Microbiology is a specialised area of biology that studies organisms that are too small to be seen without magnification. These are microorganisms, or microbes. In today's world, microbiology makes up one of the largest and most complex biological sciences because it deals with microbe-human and microbe-environmental interactions. These interactions are relevant to disease (in both plants and animals), drug therapy, immunology, genetic engineering, industry, agriculture and ecology.

Microorganisms are capable of using a wide range of substrates for metabolism and are used to produce a variety of different products which are beneficial to mankind. As a result of their adaptability, microorganisms are found in a wide range of ecological niches and can be used for a variety of research and industrial uses because of their ease of cultivation and speed of growth.

Earlier classification systems described organisms as being prokaryote or eukaryote. While the term eukaryote still applies, prokaryotes are now described as **archaea** or bacteria. The significance of this is that the three domains archaea, bacteria and eukaryotes are thought to have evolved separately from an early common ancestor.

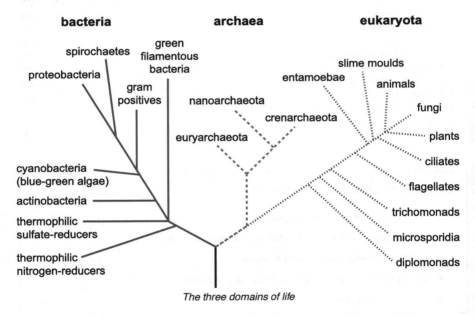

The three domains of life

TOPIC 6. ENVIRONMENTAL CONTROL OF METABOLISM

6.1.1 Archaea

Archaea share some features with both bacteria and eukaryotes, yet are significantly different to merit their own grouping.

They are single-celled and have no nucleus or organelles. There are currently four or five major sub-groups, but this may alter with further research. While outwardly appearing similar to bacteria, several of the metabolic pathways are similar to eukaryotes. They are significantly different to both in the type and composition of the lipids in their membranes. Many would have been classified as extremophiles and the properties that allowed them to exploit these niches make them of potential use to industry.

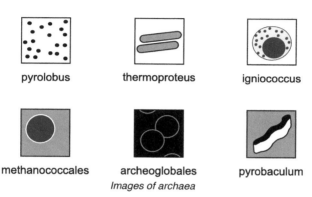

pyrolobus   thermoproteus   igniococcus

methanococcales   archeoglobales   pyrobaculum

Images of archaea

6.1.2 Bacteria

Bacteria can be classified into groups that contain organisms with similar characteristics. Bacteria are divided into three main groups. Each group is further divided until the species level is reached. The members of a bacterial species are similar to each other but can be distinguished from other species on the basis of several characteristics.

Occasionally, however, not all of the members are identical in what might be thought of as a 'pure' species culture. Different types within a species are called strains; that is, groups of different cells derived from a single cell. Strains may be identified by numbers, letters or names, for example, *E. coli* O157 or *E. coli* O111.

A variety of cell shapes (morphologies) and arrangements for bacteria exist as shown as follows.

Cocci

The number of planes in which the cells divide determines the final arrangement of the cells. Division may occur in a single plane (*Diplococcus* and *Streptococcus*), two planes (to produce tetrads), or in many planes (*Staphylococcus*).

Spirilla

There are many variations on this shape. Other shapes of bacteria do occur; *Stella*, for example, is star-shaped, and *Haloarcula* is square-shaped.

Bacilli

These rod-shaped bacteria can appear as single cells, in pairs or in chains.

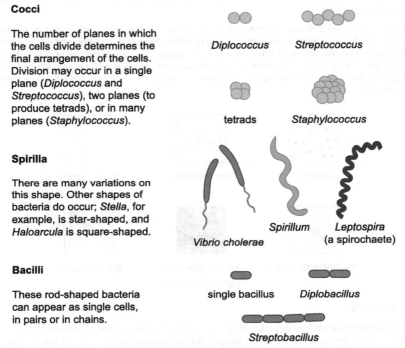

Bacterial cell shapes and arrangements

Many species of bacteria are of economic importance to humans. Bacteria are involved in the production of yoghurt, cheese, biofuels and many other products. Research into the properties of different bacterial species is also import because some cause disease. Research into bacterial species which have become resistant to antibiotics is of particular importance, i.e. MRSA.

6.1.3 Fungi

Fungi are eukaryotic cells and can be further sub-divided into yeasts and moulds. These differ with respect to their morphology. Fungi are of importance to humans because they can be both beneficial and harmful. Fungi act as decomposers, a role that is of great environmental significance.

Fungi are the major cause of plant disease. Over 5,000 species attack economically important crops and wild plants. Many animal diseases are also caused by fungi. Fungi are essential to many industrial processes that involve fermentation, such as bread, wine and beer production. They also play a major role in the manufacture of cheese and soy sauce. Fungi are essential for the commercial production of some organic acids, such as citric acid and gallic acid, and certain drugs, such as ergometrine and cortisol. They are also important in the manufacture of antibiotics (penicillin) and the immunosuppressive drug cyclosporin.

Yeasts are single cells, whereas moulds are multicellular. Generally, yeasts are larger than most bacteria and are usually oval in shape, although some may be elongated or spherical. They lack flagella and other means of locomotion. They form smooth, glistening colonies on an agar medium that are quite different from the spreading, furry, filamentous colonies formed by the moulds.

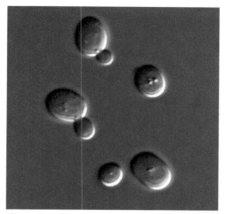

Scanning electron microscope image of yeast

6.1.4 Protozoa

Protozoa are eukaryotic cells that have a variety of shapes, such as oval, spherical or elongated. Some are polymorphic with different forms at different stages of the life cycle. Some are up to 2mm in length and, as such, are visible to the naked eye. Like animal cells, the protozoa lack cell walls, are motile at some stage of the life cycle, and are **heterotrophic**. Each individual cell is a complete organism containing all the organelles necessary to perform all the functions of an individual organism. Typical examples of protozoa are *amoeba, paramecium, euglena* and *plasmodium*.

Protozoa grow in a variety of moist habitats. They are susceptible to desiccation, so a moist environment is essential to their existence. Protozoa are predominantly free living and are found in both freshwater and marine environments. Terrestrial protozoa can be found in soil, sand and decaying matter. Many protozoa are parasites of animals or plants.

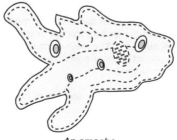

An amoeba

The cellular morphology of protozoan organisms is essentially the same as that of all eukaryotic multicellular organisms, but because all life processes occur within the single cell, there are some morphological and physiological features that are unique to protozoa.

6.1.5 Algae

Unicellular algae may be spherical, rod-shaped, club-shaped or spindle-shaped. Some are motile. Multicellular forms vary in shape and size, occurring sometimes in colonies, either as groups of single cells or a mix of different cell types with special functions. They also vary in the photosynthetic pigments present and their colour. Algae grow in a wide variety of habitats, from oceans to freshwater ponds. Algae may also be symbionts or parasitic, and they make a vital contribution to global productivity.

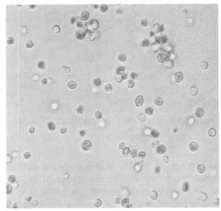

Algae

6.2 Growth of microorganisms

Microorganisms can be cultured relatively easily in a laboratory. They must be given an appropriate growth medium and the environmental conditions must be carefully controlled to ensure successful growth.

Microorganisms are grown on nutrient material called culture media. A medium is any solid or liquid preparation made specifically for the growth, storage or transport of microorganisms, which generally supplies all the essential nutrients. The medium must also be at the correct pH and the organisms have to be grown in the correct gaseous environment. The medium must be sterile before use. Two types of media are commonly used:

- **complex media** contain one or more crude sources of nutrients and their exact chemical composition and components are unknown;
- **defined media**, or synthetic media, are media in which the components of the medium are chemically known and are present in relatively pure form.

For the routine culture of microorganisms, complex media are generally used. Defined media are used in nutritional, genetic and physiological studies. Some varieties of media and the bacteria grown on them are summarised in the table below.

Type of medium	Example	Extra constituents	Bacteria grown
Complex media	nutrient agar	meat extracts, yeast extract	many bacteria will grow on this medium
Defined media	M9	M9 salts	*Escherichia coli*
Enriched media	blood agar	blood	*Streptococcus pyogenes*
Selective media	MacConkey agar	contains bile salts and crystal violet dye	inhibits the growth of Gram +ve bacteria and encourages the growth of Gram -ve bacteria

Growth media for bacteria

Microorganisms require an energy source (which may be chemical or light) and simple chemical compounds for biosynthesis. Many microorganisms can produce all of the complex molecules required for life, including amino acids for protein synthesis. Other microorganisms require more complex compounds to be added to the growth media, including vitamins and fatty acids.

In order to grow cells in culture, they must be supplied not only with the correct **nutrient medium** but also the correct environmental conditions, including:

- temperature (controlled using an incubator);
- pH (controlled by the use of buffers or addition of acid/alkali);
- gaseous environment (some microorganisms are anaerobic and will not grow in the presence of oxygen, others will require a good oxygen supply);
- light (if it is a photosynthetic microorganism).

Microorganisms are ubiquitous; any natural habitat will contain a diverse microbial population. All inanimate and living objects in the laboratory, including the atmosphere, carry large numbers of microorganisms, so special techniques are required to prevent such microorganisms from contaminating pure cultures. This is called aseptic technique and its function is three-fold:

1. to prevent contamination of cultures by unwanted organisms;

2. to prevent the organism that is being cultured from contaminating the environment (that is, the laboratory and the people in it);

3. to reduce competition with desired micro-organisms for nutrients and reduce the risk of spoilage of the product.

One way to ensure that equipment is sterile is to use heat sterilisation. During this process, all utensils (such as inoculating wires and loops) and media used to handle and grow pure cultures of microorganisms are sterilised beforehand in order to eliminate all the organisms present in or on them.

Sterilisation is normally carried out by autoclaving or by dry heat (in an oven). An autoclave is shown below. The essential part of autoclaving is that the high temperature and pressure is maintained for at least 15 minutes. This is because the spores of some bacteria are only killed at these heat and pressure intensities. Microorganisms can be grown on sterile agar or sterile broth in flasks, bottles or tubes. In each case, the vessel is topped with a metal or plastic cap, a cotton wool bung, or a foam plug designed to exclude microorganisms.

A benchtop autoclave
(http://commons.wikimedia.org/wiki/File:Laboratory_autoclave.jpg?uselang=en-gb by **Nadina Wiórkiewicz (pl.wiki:** http://en.wikipedia.org/wiki/pl:user:Nadine90, **commons:** http://commons.wiki media.org/wiki/User:Nadine90), **licensed under** http://creativecommons.org/licenses/by-sa/3.0 **via** http://commons.wikimedia.org/)

6.3 Patterns of growth

Under ideal conditions, some species of bacteria are capable of doubling in number every 20 minutes. If you were trying to plot bacterial growth on normal graph paper, you would either run out of space very quickly on the x axis, or the scale would be so reduced it would make plotting or reading with any accuracy almost impossible. The solution is to use semi-logarithmic graph paper. This has been printed in a specific way to allow data which has a very wide range to be plotted. This allows exponential relationships to be depicted. In microbial cases, time is usually plotted as the independent variable (x axis) and logarithmic growth of bacteria is plotted as the dependent variable (y axis).

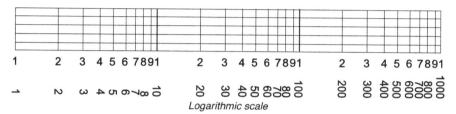

The extent of growth of a microbial culture can be estimated by taking samples from the culture at certain time points and counting the number of cells present at each one. A count is also made at the time of inoculation so that the initial concentration of cells is known. The results of the timed samplings will give enough information to construct a growth curve.

The four main stages in growth are:

1. **lag phase** - where microorganisms adjust to the conditions of the culture by producing enzymes that metabolise the available substrates;

2. **exponential (logarithmic) phase** - during this phase the rate of growth is at its highest due to plentiful nutrients;

3. **stationary phase** - occurs due to the nutrients in the culture media becoming depleted and the production of toxic **metabolites**. Secondary metabolites are also produced, such as antibiotics. In the wild these metabolites confer an ecological advantage by allowing the micro-organisms which produce them to outcompete other micro-organisms;

4. **death phase** - occurs due to the toxic accumulation of metabolites or the lack of nutrients in the culture.

Note the use of a logarithmic scale on the vertical axis in the following graph.

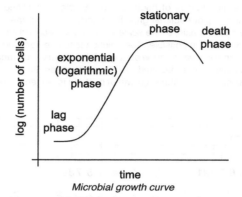
Microbial growth curve

This following illustration shows the change in turbidity (optical density) of a culture as it grows.

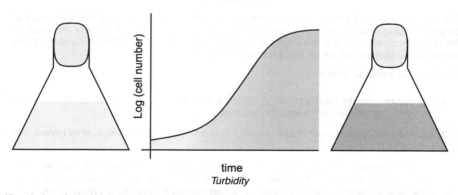
Turbidity

Knowledge of microbial growth rates is essential to microbiologists because growth rate studies are relevant to fundamental research and also to applied situations, such as the industrial production of microorganisms. A knowledge of growth rates allows scientists to predict, and possibly control, the growth of any unicellular microbial species.

During the exponential phase of growth, a microorganism is dividing at constant intervals. The population will therefore double during a certain period of time, called the generation time. Not all species of microorganism have the same generation time. For example, *E. coli* grown in an appropriate medium has a generation time as low as 15 minutes. *Mycobacterium tuberculosis*, on the other hand, has a generation time of 13-15 hours. In addition, generation time for a species will vary if growth conditions change. For example, *E. coli* will take much longer to divide than in the example above if it is grown in a nutritionally poor medium. Examples of typical generation times for different types of microorganism are given in the following table.

Microorganism	Generation time (hours)
Bacteria	
Escherichia coli	0.35
Bacillus subtilis	0.43
Clostridium botulinum	0.58
Algae	
Chlorella pyrenoidosa	7.75
Skeletonema costatum	13.1
Euglena gracilis	10.9
Protozoa	
Paramecium caudatum	10.4
Fungi	
Saccharomyces cerevisiae	2.0

Generation times for microorganisms

When cell counts are performed, they fall into one of two categories: total cell count or viable cell count. A total cell count involves a direct counting of the cells in culture whether dead or alive. A viable cell count is a method where only live cells are counted. Only viable cell counts show a death phase where cell numbers are decreasing.

Bacterial culture growth phases: Questions

This is an analysis activity. Each letter represents a different phase of growth for a bacterial culture. Study the graph and then answer the questions that follow.

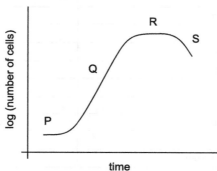

Bacterial culture growth phases

Q1: In which phase are the bacteria doubling at a constant rate?

a) P
b) Q
c) R
d) S

....

Q2: In which phase does bacterial cell division equal bacterial death?

a) P
b) Q
c) R
d) S

....

Q3: In which phase are the bacteria metabolically active but not dividing?

a) P
b) Q
c) R
d) S

....

Q4: In which phase does bacterial cell death exceed cell division?

a) P
b) Q
c) R
d) S

6.4 Penicillin production

Many antibiotics are produced by microorganisms, predominantly by the actinomycetes *Streptomyces* spp. and some filamentous fungi. Environmental control is crucial to the production of antibiotics such as penicillin.

Penicillium chrysogenum is the fungus that produces penicillin. It is the control of the composition of the medium in which the *Penicillium* is grown that allows high yields of the product to be obtained. The organism is grown in stirred fermenters but the rapid growth of the cells, with glucose as a carbon source, does not necessarily lead to maximum antibiotic yields. The greatest yield occurs when the medium is provided with a combination of lactose, as a carbon source, with a limited nitrogen availability that stops growth. However, the same result can be achieved by using a slow continuous feed of glucose and this is the method used today.

It can be seen, therefore, that the manipulation of the growth medium or the organism itself can result in high yields of the product.

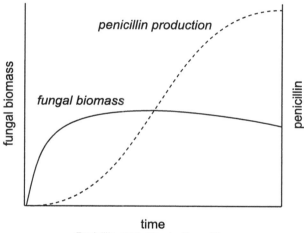

time
Penicillin production by Penecillium

6.5 Learning points

Summary

- Microbiology is the study of organisms that are too small to be seen by the naked eye.
- Microbes can be found in all three domains: archaea, bacteria and eukaryotes.
- As a result of their adaptability, microorganisms are found in a wide range of ecological niches, and can be used for a variety of research and industrial uses because of their ease of cultivation and speed of growth.
- Microorganisms require an energy source (chemical or light) and simple chemical compounds for biosynthesis.
- Many microorganisms can produce all the complex molecules they require. Other microorganisms require more complex compounds to be added to the growth media.
- Culture conditions include sterility, to eliminate any effects of contaminating microorganisms, control of temperature in an incubator, control of oxygen levels by aeration and control of pH by buffers or the addition of acid or alkali.
- Microbes grow in vast numbers; plotting growth requires the use of semi logarithmic scales.
- The growth cycles consists of distinct phases: lag, log, stationary and death.
- During the lag phase, microorganisms adjust to the conditions of the culture by producing enzymes that metabolise the available substrates.
- During the log phase, the rate of growth is at its highest due to plentiful nutrients.
- The stationary phase occurs due to the nutrients in the culture media becoming depleted and the production of toxic metabolites. Secondary metabolites are also produced, such as antibiotics. In the wild these metabolites confer an ecological advantage by allowing the micro-organisms which produce them to outcompete other micro-organisms.
- During the death phase, a lack of substrate and the toxic accumulation of metabolites cause death of cells.
- Secondary metabolites have no direct relationship to the synthesis of cell materials and normal growth; they may confer an ecological advantage.

6.6 Extended response question

The activity which follows presents an extended response question similar to the style that you will encounter in the examination.

You should have a good understanding of the patterns of growth shown by microbes before attempting the question.

You should give your completed answer to your teacher or tutor for marking, or try to mark it yourself using the suggested marking scheme.

Extended response question: The patterns of growth shown by microbes

Give a brief description of the patterns of growth shown by microbes. *(6 marks)*

6.7 Extension materials

The material in this section is not examinable. It includes information which will widen your appreciation of this section of work.

Extension materials: Diauxic growth and catabolite repression

The mechanism of diauxic growth can be conveniently explained using the growth of *Escherichia coli* (*E. coli*) as an example. In a culture medium there may only be one source of sugar, such as glucose, for bacterial metabolism. However, when there is more than one sugar available, which sugar does the bacterium metabolise first or are they metabolised together? These are the questions that were asked in the 1940s by the distinguished scientist Jacques Monod.

The answer is that bacteria use sophisticated mechanisms to use all of the available sugar efficiently. In a culture medium containing both glucose and lactose, for example, the glucose is always metabolised first. This is because it is metabolised much more efficiently than lactose. After the available glucose has been used up, the lactose is metabolised. This results in a two-step growth curve and this form of growth is called diauxic growth.

How does this happen? Glucose prevents growth on lactose by acting as a repressor in the synthesis of the enzymes of the lactose operon. This is called catabolite repression. It is known that glucose represses the synthesis of a large number of enzymes in many pathways and so it is considered to be a global repressor.

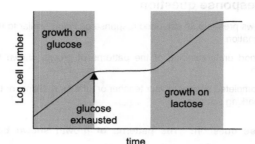

Diauxic growth by E. coli when grown on a mixture of glucose and lactose

The main features of glucose repression involve a reduction in the level of cyclic AMP (cAMP), and the inhibition of an enzyme (adenylate cyclase) involved in cAMP synthesis. cAMP forms a complex with an activator protein (called CAP, for cAMP Activated Protein) that binds to the promoter of glucose-repressible genes. When glucose is absent, cAMP levels are high, cAMP is bound to CAP, and there is no glucose repression of gene expression. When glucose is present, cAMP levels are low, CAP is inactive, and RNA polymerase does not readily bind to the promoter of glucose-repressible genes. Consequently, the level of gene expression is very low.

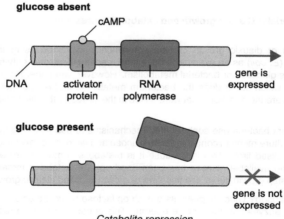

Catabolite repression

Why is catabolite repression important? It ensures that a cell uses the most readily metabolised source of energy in its environment. This will therefore support the most rapid growth of the cell. Since glucose is metabolised faster than lactose in *E. coli*, it is used first. Lactose will be metabolised only when glucose is depleted. This regulatory mechanism has one purpose and that is to allow bacteria to reproduce at their maximum rate in any environment.

6.8 End of topic test

End of Topic 7 test Go online

Q5: Living organisms are classified under three domains. Name these domains.

....

Q6: List two environmental conditions that need to be controlled for microbes to grow successfully.

....

Q7: The following graph shows a typical growth pattern for micro-organisms in culture.

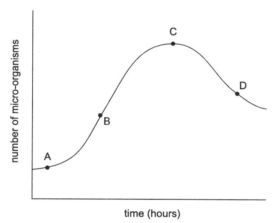

At which lettered stage does the death rate exceed the rate of cell division?

....

Q8: Arrange the processes in a growth curve into the correct order.

1. Lag phase -
2. Log phase -
3. Stationary phase -
4. Death phase -

Processes:

- decline of population due to exhaustion of nutrients and build-up of toxins;
- organisms acclimatising to their environment;
- organisms growing exponentially;
- organisms likely to be producing secondary metabolites.

....

Q9: The following diagram shows a bacterial cell that has been magnified 800 times.

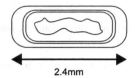

2.4mm

What is the actual length of the cell in μm?

a) 0.003
b) 0.03
c) 0.3
d) 3.0

....

Q10: Bacterial cells were exposed to disinfectant for increasing lengths of time to determine the number of live cells left after treatment. The following graph shows the number of bacterial cells which survived.

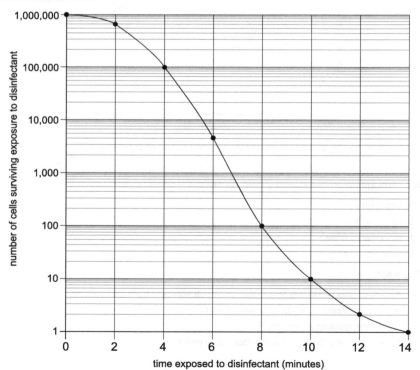

How many cells survive after 6 minutes?

a) 1300
b) 4000
c) 5000
d) 5500

Unit 2 Topic 7

Genetic control of metabolism

Contents

- 7.1 Improving wild strains of microorganisms 221
- 7.2 Mutagenesis 221
- 7.3 Selective breeding 222
- 7.4 Recombinant DNA 222
- 7.5 Recombinant DNA technology 223
- 7.6 Bovine somatotrophin (BST) 226
- 7.7 Learning points 228
- 7.8 Extension materials 229
- 7.9 End of topic test 230

Prerequisites

You should already know that:

- a mutation is a random change to genetic material;
- mutations may be neutral, or confer an advantage or a disadvantage;
- mutations are spontaneous and are the only source of new alleles;
- environmental factors, such as radiation and chemicals, can increase the rate of mutation.

Learning objective

By the end of this topic, you should be able to:

- explain that wild strains of micro-organisms can be improved by mutagenesis, or recombinant DNA technology;
- explain that mutations can be induced by radiation or chemicals;
- explain that plant and animal genes can be transferred to microbes, causing them to express the donor gene products (proteins);
- know that yields can be increased by gene manipulation;
- know that safety mechanisms can also be introduced, such as introducing genes which prevent the growth of genetically modified organisms outside the laboratory environment;
- explain that extra chromosomal DNA, such as plasmids, can be transferred to microorganisms which contain a variety of genes with specific functions;
- describe the roles of restriction endonuclease and ligase enzymes;
- know that some plant and animal proteins may be produced more successfully in a recombinant yeast cell, rather than a recombinant bacterial cell.

TOPIC 7. GENETIC CONTROL OF METABOLISM

7.1 Improving wild strains of microorganisms

When working with microorganisms in a lab, it may be necessary to improve the strain you are using, for example by enabling it to produce a desired product in large quantities. This can be achieved by one of two different methods:

- mutagenesis;
- recombinant DNA.

7.2 Mutagenesis

Mutations can, on very rare occasions, give rise to improved forms of any organism. The same is true for microorganisms. The major difference is that the generation time for many microbes is very short when compared to humans, for example. This means that, while mutation rates may be no greater, in a relatively short period of time in human terms, large numbers of mutants may arise within a population of microorganisms. When coupled to the vast numbers in which they are found, microbe mutants can be induced and isolated quite easily.

Mutation rates in humans and bacteria

Normal mutation rate is given as approximately one in one hundred million. If the population of the USA is roughly 310 million, this would give rise to only three people having a mutation for a particular trait. Assuming a world population of seven thousand million (7 billion) and the same rate of mutation, this would suggest only 70 people in the world would exhibit this mutation. However, within the human gut, any one millilitre may contain 10 billion bacteria. This greatly increases the number of mutations for any given characteristic simply because the numbers of bacteria are so much greater.

Mutations can occur naturally and are mostly harmful to an organism. Those which are beneficial are very rare and, even more rarely, they can revert to **wild-type**, i.e. normal. Mutation rates are greatly increased by exposing the DNA to certain mutagenic agents. These have the effect of altering the DNA sequence (the genome) and, as a result, there may be a change in the proteins produced.

The process of creating mutants is known as mutagenesis. By exposing microorganisms to mutagens, for example UV light, the rate of mutagenesis can be increased. Scientists can expose microorganisms to various mutagens and then screen them for a specific desired characteristic, for example the ability to grow on cheap nutrients. These mutants can then be cultured and used in the lab.

Although this seems a relatively straightforward way to produce a desirable characteristic in a microorganism, mutant strains tend to be genetically unstable. This means they are likely to revert back to their wild type (normal) phenotype.

© HERIOT-WATT UNIVERSITY

7.3 Selective breeding

Some microorganisms (such as fungi) are capable of both asexual and sexual reproduction. Asexual reproduction produces offspring that are identical to the parent which produced them. Sexual reproduction produces offspring which have different characteristics from their parents, in other words they show variation.

Selective breeding involves sexual reproduction between two strains of microorganism, each with a desirable characteristic. The aim of selective breeding is to produce offspring which have a new genotype and that show both desirable characteristics. Selective breeding can take a long time to achieve the desired result as it is also likely that undesirable characteristics will combine and show up in the phenotype of the offspring.

7.4 Recombinant DNA

For the greater part of their existence, bacteria reproduce asexually by binary fission, resulting in clones. However, there are occasions when genetic material is exchanged horizontally, i.e. between members of the same generation. This can result in new strains of bacteria which show variation from the original strain.

Some bacteria are capable of transferring **plasmids** or pieces of chromosomal DNA between one another, whilst others are capable of taking up DNA from their environment and incorporating it into their genome. Scientists can take advantage of this to attempt to produce new strains of bacteria with desirable characteristics.

TOPIC 7. GENETIC CONTROL OF METABOLISM

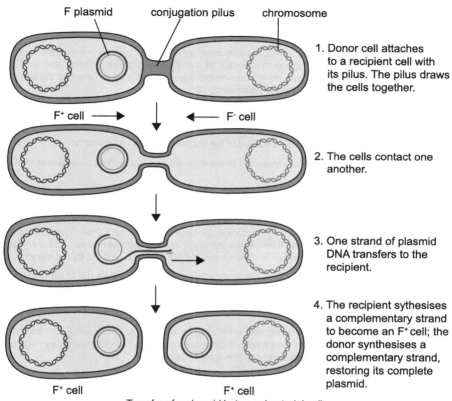

Transfer of a plasmid between bacterial cells

7.5 Recombinant DNA technology

Recombinant DNA technology refers to the ability of scientists to manipulate the genome of an organism by introducing new gene sequences. New gene sequences can be transferred between individuals of the same species or from one species to another. Using recombinant DNA technology, it is possible to artificially improve a strain of microorganism, for example allowing scientists to increase the yield of a desired product by introducing genes that remove inhibitory controls or amplify specific metabolic steps in a pathway. As a safety mechanism, genes are often introduced that prevent the survival of the microorganism in an external environment.

In order to introduce a gene into an organism, scientists must use a vector. A vector is a DNA molecule used to carry foreign genetic information into another cell and both **plasmids** and artificial chromosomes are used as vectors during recombinant DNA technology. Recombinant plasmids and artificial chromosomes share many features, but plasmids are only capable of carrying relatively small quantities of DNA whereas artificial chromosomes can carry longer DNA sequences.

Recombinant plasmids and artificial chromosomes contain the following features which allow them to operate effectively:

- selectable markers - such as antibiotic resistance genes protect the micro-organism from a selective agent (antibiotic) that would normally kill it or prevent it growing. Selectable marker genes present in the vector ensure that only micro-organisms that have taken up the vector grow in the presence of the selective agent (antibiotic);
- restriction sites - contain target sequences of DNA where specific restriction endonucleases cut;
- origin of replication - allows self-replication of the plasmid/artificial chromosome;
- regulatory sequences - control expression of the inserted gene as well as other genes found on the vector.

The manipulation of DNA involves the use of two types of enzyme: **restriction endonuclease** and **ligase**.

Restriction endonucleases are enzymes that cut specific target sequences of DNA. Typically, there is a sequence of four to six nucleotide bases which the enzyme will recognise and cut. There will be complementary sequences opposite on the other strand of DNA, which is also cut by the enzyme.

For example, the enzyme EcoR1 binds at the sequence *GAATTC* and breaks the chain between the G and first A, leaving a small section of single-stranded DNA *AATTC* - a so called 'sticky end'. It is called this because the same happens on the opposite strand of the same double-stranded DNA molecule.

These ends could re-join by base-pairing with each other or with similar sticky ends from another source, for example a plasmid as shown below. If base-pairing occurs between two sticky ends of DNA, the backbone can be re-joined by an enzyme called ligase.

Recombinant DNA: Steps

The following shows the steps involved in recombinant DNA.

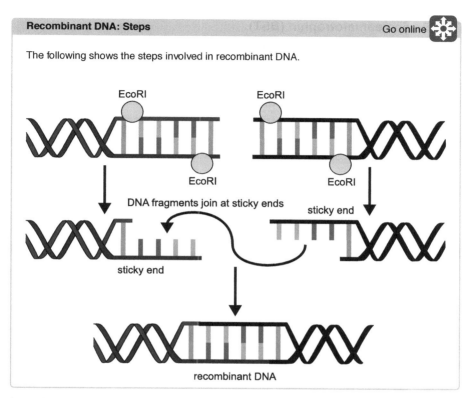

It can be difficult to express animal or plant (eukaryotic) genes successfully in bacterial cells. Bacterial proteins do not undergo the same post-transcriptional and post-translational modifications which eukaryotic proteins do. Expression of plant or animal genes in bacteria may result in polypeptides that are folded incorrectly and are therefore non-functional. In cases where this happens, the eukaryotic proteins may be produced more successfully in a recombinant yeast cell.

7.6 Bovine somatotrophin (BST)

Increasingly, farm animals are being treated with products made by recombinant DNA technology. These include vaccines, antibodies and growth hormones. In the USA, dairy cattle are routinely injected with a growth hormone called bovine somatotrophin, or BST, which increases milk production by about 10%. BST also improves weight gain in beef cattle.

To produce pure BST from the cloned gene, the following steps were taken:

- a **restriction endonuclease** enzyme was used to isolate the BST gene on a small fragment of DNA;
- **plasmid** DNA was isolated from *E. coli*;
- the same restriction endonuclease enzyme that was used to cut out the BST gene was also used to cut the plasmid DNA;
- the BST gene was inserted, or ligated, into the bacterial plasmid;
- the recombinant plasmid was used to transform *E. coli* cells.

The production of large quantities of BST requires the following:

- propagation of the transformed *E. coli* expressing the BST gene in large vats containing nutrient broth;
- purification of the BST secreted by the *E. coli* into the nutrient broth.

Some countries (including the UK) continue to ban the import of milk produced by cows injected with BST. The concern is that the milk could contain minute quantities of BST and may have potential side-effects.

Bovine somatotrophin: Questions   Go online

Q1: Bacterial plasmids are described as vectors. What is meant by the term 'vector'?

....

Q2: What is a 'recombinant plasmid'?

....

Q3: _____ are enzymes that cut up DNA into fragments.

a) Ligases
b) Restriction endonuclease enzymes

....

Q4: Bacterial plasmids are cut open using _____.

a) Ligases
b) Restriction endonuclease enzymes

....

TOPIC 7. GENETIC CONTROL OF METABOLISM

Q5: _____ are enzymes that are used to seal DNA fragments into plasmids.

a) Ligases
b) Restriction endonuclease enzymes

Q6: Complete the diagram that shows the cloning of the bovine somatotrophin gene using the labels provided.

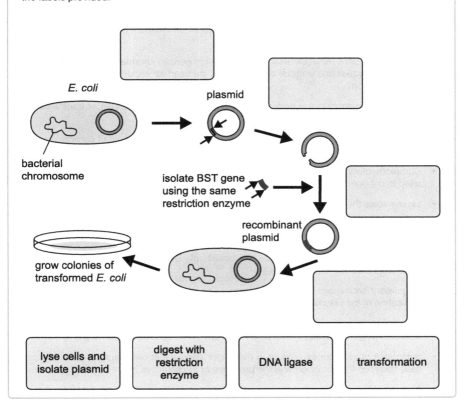

7.7 Learning points

Summary

- Wild strains of micro-organisms can be improved by mutagenesis, or recombinant DNA technology.
- Exposure to UV light and other forms of radiation or mutagenic chemicals results in mutations, some of which may produce an improved strain of micro-organism.
- Recombinant DNA technology involves the use of recombinant plasmids and artificial chromosomes as vectors.
- A vector is a DNA molecule used to carry foreign genetic information into another cell and both plasmids and artificial chromosomes are used as vectors during recombinant DNA technology.
- Artificial chromosomes are preferable to plasmids as vectors when larger fragments of foreign DNA are required to be inserted.
- Restriction endonucleases cut open plasmids and specific genes out of chromosomes, leaving sticky ends.
- Complementary sticky ends are produced when the same restriction endonuclease is used to cut open the plasmid and the gene from the chromosome.
- Ligase seals the gene into the plasmid.
- Recombinant plasmids and artificial chromosomes contain restriction sites, regulatory sequences, an origin of replication and selectable markers.
- Restriction sites contain target sequences of DNA where specific restriction endonucleases cut.
- Regulatory sequences control gene expression and origin of replication allows self-replication of the plasmid/artificial chromosome.
- Selectable markers such as antibiotic resistance genes protect the micro-organism from a selective agent (antibiotic) that would normally kill it or prevent it growing.
- Selectable marker genes present in the vector ensure that only micro-organisms that have taken up the vector grow in the presence of the selective agent (antibiotic).
- As a safety mechanism, genes are often introduced that prevent the survival of the micro-organism in an external environment.
- Recombinant yeast cells may be used as plant or animal recombinant DNA expressed in bacteria may result in polypeptides being incorrectly folded.

7.8 Extension materials

The material in this section is not examinable. It includes information which will widen your appreciation of this section of work.

Extension materials: Restriction endonuclease EcoR1

In 1973, Cohen and Boyer demonstrated that frog DNA which coded for ribosomal RNA could be transferred into a bacterium. They constructed a plasmid (pSC101) which contained two specific features. One was a single site for the attachment of the restriction endonuclease EcoR1. The other was a gene conferring resistance to the antibiotic tetracycline.

By incubating the frog DNA with EcoR1, it was cut at specific points into fragments with sticky ends. Similarly, the plasmids were digested by the EcoR1, revealing the complementary sticky ends. The frog DNA fragments were mixed with the opened plasmids and recombination took place.

Various fragments of frog DNA were incorporated into the plasmids and some would include the ribosomal RNA gene. The enzyme ligase was introduced to join the phosphodiester bonds and link the DNA into the plasmids, which were in turn reintroduced to *E.coli* bacteria by transformation. These *E.coli* were sensitive to tetracycline and, as a result, only those bacteria carrying the plasmid with gene for resistance would survive when grown in a medium containing the antibiotic. Of the surviving colonies, those containing rRNA gene were selected.

7.9 End of topic test

End of Topic 8 test Go online

Q7: Most mutations are beneficial. True or false?

Q8: Mutagenic agents can be roughly placed into two groups. Name them.

Q9: What name is given to an agent which can carry DNA into a cell?

Q10: In recombinant DNA technology, name the enzyme which cuts DNA leaving sticky ends.

Q11: Name the enzyme which joins sticky ends of DNA together.

Q12: Describe one feature which must be present on a plasmid to allow it to act as an effective vector.

Q13: If an animal protein is expressed in a bacterial cell, what problems could arise with its production?

Q14: If an animal protein is not expressed successfully in a bacterial cell, what type of cell could it be expressed in more successfully?

Unit 2 Topic 8
End of unit test

End of Unit 2 test

Go online

Metabolic pathways

Q1: Metabolic pathways are regulated by _____.

Q2: Reactions which build up large molecules from small molecules are known as _____.

Q3: Reactions which break down large molecules into smaller molecules are known as _____.

The enzyme shown in the following diagram controls one step of a metabolic pathway.

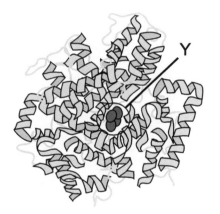

Q4: In the diagram, what is represented by the letter Y?

Q5: Explain the term 'induced fit'.

Q6: Explain the difference between a competitive inhibitor and non-competitive inhibitor.

Q7: The following diagram shows the effects of enzyme inhibition on reaction rate. The curve labelled A shows the rate of reaction in the absence of an inhibitor.

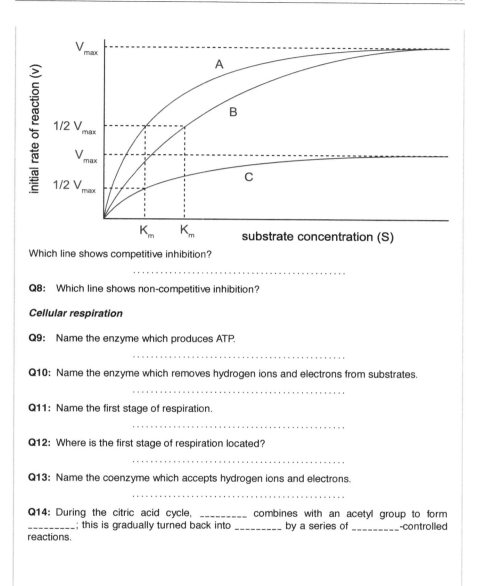

Which line shows competitive inhibition?

Q8: Which line shows non-competitive inhibition?

Cellular respiration

Q9: Name the enzyme which produces ATP.

Q10: Name the enzyme which removes hydrogen ions and electrons from substrates.

Q11: Name the first stage of respiration.

Q12: Where is the first stage of respiration located?

Q13: Name the coenzyme which accepts hydrogen ions and electrons.

Q14: During the citric acid cycle, _____ combines with an acetyl group to form _____; this is gradually turned back into _____ by a series of _____-controlled reactions.

Metabolic rate

Q15: Which of the following measurements do not allow metabolic rate to be calculated?

a) Calorie intake
b) Carbon dioxide production
c) Heat generation
d) Oxygen consumption

....

Q16: The following diagram shows the structure of three different types of heart.

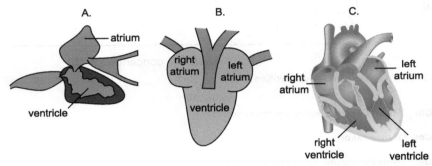

What is the correct order of organisms to match with the hearts in the order as labelled A-C?

a) Amphibian, fish, mammal
b) Amphibian, mammal, fish
c) Fish, amphibian, mammal
d) Mammal, fish, amphibian

TOPIC 8. END OF UNIT TEST

Metabolism in conformers and regulators

Q17: Name an external abiotic factor which can affect the ability of an organism to maintain its metabolic rate.

....

Q18: The following diagram shows the control of body temperature in a mammal.

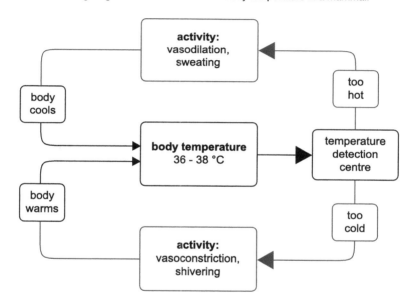

What is the location of the temperature detection centre?

....

Q19: What name is given to the corrective mechanism shown in the diagram?

....

Q20: Why is it important for mammals to control their body temperature?

....

Q21: A _____ has an internal environment which is dependent on the external environment. They have _____ metabolic costs and inhabit a _____ range of ecological niches.

....

Q22: A _____ uses energy to control its internal environment. They have _____ metabolic costs and inhabit a _____ range of ecological niches.

Maintaining metabolism during environmental change

Q23: Some animals avoid extreme or hostile environments by _____. *(Choose from 'hibernation' or 'migration'.)*

....

Q24: Successful migration depends on a combination of _____ and _____ behaviours.

....

Q25: What name is given to dormancy which occurs after the onset of adverse environmental conditions?

....

Q26: What name is given to dormancy which occurs before the onset of adverse environmental conditions?

Environmental control of metabolism

Q27: Name the three domains of life.

....

Q28: As a result of their adaptability, microorganisms are found in a _____ range of ecological niches.

....

Q29: List two environmental conditions that need to be controlled for microbes to grow successfully.

....

Q30: The following diagram shows a typical growth curve.

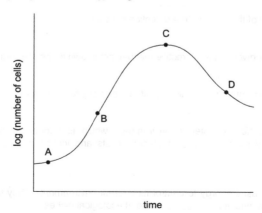

Name each of the stages shown in the graph.

TOPIC 8. END OF UNIT TEST

Genetic control of metabolism

Q31: Name one way by which wild strains of microorganisms can be improved.

....

Q32: Some bacteria can transfer plasmids or pieces of chromosomal DNA to each other, resulting in the production of new strains of bacteria. What is this type of transfer known as?

....

Q33: In recombinant DNA technology, name the enzyme which cuts DNA leaving sticky ends.

....

Q34: Name the enzyme which joins sticky ends of DNA together.

....

Q35: When working with _____ in a lab, as a safety mechanism, genes are often introduced that prevent the survival of the _____ in an _____ environment.

....

Q36: Plant or animal recombinant DNA in bacteria may result in polypeptides that are folded incorrectly. How can these polypeptides be expressed more successfully?

Problem solving

The tables below show the growth of *E. coli* in nutrient broth containing different sugar substrates. Nutrient broth was prepared with one of two different sugars, inoculated with *E. coli*, and the number of viable cells was calculated every hour for 12 hours.

Time (hours)	Viable cell number (millions per cm^3)
0	12
1	12
2	12
3	17
4	44
5	75
6	91
7	92
8	93
9	94
10	94
11	94
12	94

Table 1
Nutrient broth containing glucose

Time (hours)	Viable cell number (millions per cm^3)
0	10
1	10
2	10
3	10
4	10
5	12
6	34
7	60
8	87
9	91
10	92
11	92
12	92

Table 2
Nutrient broth containing lactose

Q37: Give a reason why it is important to use separate syringes when preparing the nutrient broth containing different sugars.

....

Q38: Name one factor which must be kept constant when setting up this experiment.

TOPIC 8. END OF UNIT TEST

Q39: Complete the graph and key to show the results for *E. coli* grown in nutrient broth containing lactose.

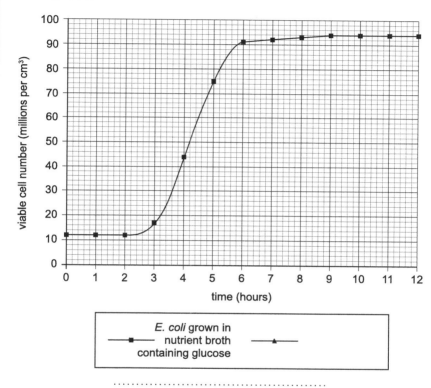

Q40: Using viable cell numbers gives more reliable results than using total cell numbers. Explain why.

Q41: How much longer does it take the culture to reach 60 million viable cells per cm^3 when using lactose as a respiratory substrate rather than glucose?

Q42: In terms of phases of growth, give two conclusions from the results of this experiment.

Unit 3: Sustainability and Interdependence

1	**Food supply**	**245**
1.1	Food security	247
1.2	Agricultural production - Food production and photosynthesis	248
1.3	Agricultural production - Food production and trophic levels	251
1.4	Agricultural production - Efficient food production	254
1.5	Learning points	255
1.6	Extension materials	256
1.7	End of topic test	257
2	**Plant growth and productivity**	**261**
2.1	Photosynthesis and energy capture	263
2.2	Photosynthetic pigments	264
2.3	The spectrum of light	265
2.4	Absorption spectrum	266
2.5	Action spectrum	267
2.6	First stage of photosynthesis: The light-dependent stage	269
2.7	The second stage of photosynthesis: The carbon fixation stage	271
2.8	Learning points	273
2.9	Extended response question	274
2.10	End of topic test	275
3	**Plant and animal breeding**	**283**
3.1	Introduction	285
3.2	Field trials	286
3.3	Inbreeding	287
3.4	Cross breeding and F1 hybrids	288
3.5	F1 Hybrids	289
3.6	Genetic technology	289
3.7	Learning points	291
3.8	Extension materials	293
3.9	End of topic test	294

4 Crop protection — 297
- 4.1 Introduction — 299
- 4.2 Weeds, pests and diseases — 300
- 4.3 Control of weeds, pests and diseases — 301
- 4.4 Problems with pesticides — 303
- 4.5 Biological control and integrated pest management — 306
- 4.6 Learning points — 308
- 4.7 End of topic test — 310

5 Animal welfare — 313
- 5.1 Animal welfare — 314
- 5.2 Behavioural indicators of poor welfare — 316
- 5.3 Learning points — 318
- 5.4 End of topic test — 318

6 Symbiosis — 321
- 6.1 Symbiosis — 323
- 6.2 Parasitism — 324
- 6.3 Malaria — 325
- 6.4 Mutualism — 326
- 6.5 Learning points — 328
- 6.6 Extended response question — 328
- 6.7 End of topic test — 329

7 Social behaviour — 333
- 7.1 Social behaviour — 335
- 7.2 Social hierarchy — 335
- 7.3 Cooperative hunting — 338
- 7.4 Social mechanisms for defence — 339
- 7.5 Altruism and kin selection — 340
- 7.6 Social insects — 344
- 7.7 Primate behaviour — 345
- 7.8 Learning points — 346
- 7.9 Extended response question — 348
- 7.10 End of topic test — 349

UNIT 3: SUSTAINABILITY AND INTERDEPENDENCE

8 Components of biodiversity **353**
  8.1 Introduction 355
  8.2 Genetic diversity 356
  8.3 Species diversity 357
  8.4 Ecosystem diversity 357
  8.5 Learning points 358
  8.6 Extended response question 359
  8.7 End of topic test 359

9 Threats to biodiversity **361**
  9.1 Overexploitation 363
  9.2 The impact of habitat loss 365
  9.3 Introduced, naturalised and invasive species 368
  9.4 Learning points 369
  9.5 Extended response question 370
  9.6 End of topic test 371

10 End of unit test **375**

Unit 3 Topic 1

Food supply

Contents

1.1 Food security 247
1.2 Agricultural production - Food production and photosynthesis 248
1.3 Agricultural production - Food production and trophic levels 251
1.4 Agricultural production - Efficient food production 254
1.5 Learning points 255
1.6 Extension materials 256
1.7 End of topic test 257

Prerequisites

You should already know that:

- fertilisers and pesticides can be used to increase crop yield;
- at each level in a food chain 90% of energy is lost as heat, movement or undigested materials.

Learning objective

By the end of this topic, you should be able to:

- explain the need to increase global food production;
- describe the role of food security on a global market;
- describe the result of increase in human population on demand for increased food production;
- outline the need for sustainable food production;
- describe the need to ensure that food production does not degrade the natural resources on which agriculture depends;
- describe how all food production is ultimately dependent on photosynthesis;
- apply the knowledge that a small number of plant crops produce most human foods;
- describe the fact that if the area to grow crops is limited, increased food production will depend on other factors that control plant growth;
- describe how control of plant growth can be influenced by the breeding of high yielding cultivars;
- describe how protecting crops from pests, disease and competition can lead to increase in growth;
- outline why livestock produce less food per unit area than plant crops;
- outline the loss of energy between trophic levels;
- describe how livestock production may be possible in managed and wild habitats unsuitable for cultivation of crops.

TOPIC 1. FOOD SUPPLY

1.1 Food security

Food security is the ability of human populations to produce food of sufficient quality and quantity. Such conditions for food security can be assessed on any scale, from a single household to a global scale.

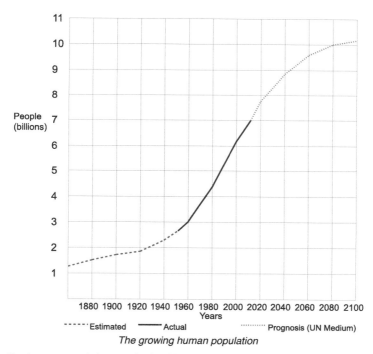

The growing human population

In 2011, the human population reached 7 billion. It is estimated that by 2050 this figure will increase to 9.4 billion. Sustaining this increasing population will not be possible without a change in agricultural practices.

There is also a demand that food production is sustainable and does not degrade the natural resources upon which agriculture depends. The Food and Agricultural Organization of the United Nations conclusion is that global food production must rise by 70% by 2050 to cater for growth in the world population of more than 30%.

Food production should be **sustainable**. Sustainability in food production can be defined as the ability of food systems to keep production and distribution going continuously without environmental degradation. It implies the ability to sustain the growth of food production to meet the demand for food in the future.

Sustainable food should be produced, processed and traded in ways that:

- contribute to thriving local economies and sustainable livelihoods - both in the UK and, in the case of imported products, in producer countries;
- protect the diversity of the environment for both plants and animals (and the welfare of farmed and wild species), and avoid damaging natural resources and contributing to climate change;
- provide benefits for society, such as good quality food, safe and healthy products, and educational opportunities.

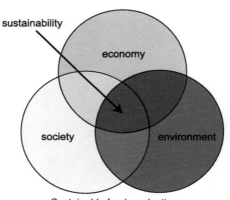

Sustainable food production

1.2 Agricultural production - Food production and photosynthesis

All food production is dependent ultimately upon photosynthesis.

Photosynthesis is a chemical process that occurs in green plants which traps light energy and converts carbon dioxide and water into organic compounds, especially sugars, using the energy from sunlight and photosynthetic pigments in the green plant.

The summary equation for photosynthesis can be written as follows:

$$6CO_2 + 6H_2O \xrightarrow{light} C_6H_{12}O_6 + 6O_2$$

carbon dioxide + water → sugar + oxygen

Starch is produced after photosynthesis when large numbers of sugar units join together. Starch is produced by all green plants as an energy store. It is the most common carbohydrate in the human diet and is contained in large amounts in such staple foods as potatoes, wheat, maize (corn) and rice. Most human food comes from a small number of plant crops (cereals, potato, roots, and **legumes**) which contain starch and other food groups.

TOPIC 1. FOOD SUPPLY

If the area to grow crops is limited, increased food production will depend on factors that control plant growth. For example the breeding of higher yielding **cultivars**. This means producing plants with an increased yield, disease/pest resistance, higher nutritional values, physical characteristics suited to rearing and harvesting or the ability to thrive in particular environmental conditions.

Methods of protecting crops from pests and diseases (for example through the use of pesticides) and reducing **competition** (for example through the use of herbicides) may also help to increase food production.

Agricultural production - Plant growth: How it works

Go online

When light from the sun shines on a plant leaf, some of it is absorbed by special pigments (chemicals) in the leaf. These pigments use the energy from the sun to produce food in the process of photosynthesis. Photosynthesis takes place primarily in plant leaves and little to none occurs in plant stems. The process of photosynthesis takes place in chloroplasts in the leaf where photosynthetic pigments are located.

The most important photosynthetic pigment is chlorophyll. The chlorophyll molecules trap the energy from light to drive a series of chemical reactions. In photosynthesis, carbon dioxide from the atmosphere and water are converted into organic compounds (especially sugars) along with the release of oxygen gas as a waste product.

The sugars and other compounds produced from photosynthesis are used for plant growth and other essential metabolic processes in plants. In addition to maintaining normal levels of oxygen, photosynthesis is the source of energy for nearly all life on earth. The production of food for plants to allow them to grow is referred to as primary **productivity**. This source of energy can be passed onto animals when they consume plant material in their food. Therefore, photosynthesis is ultimately the source of almost all the food on earth.

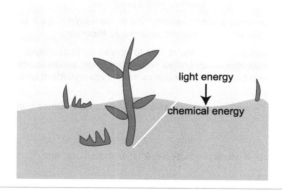

Agricultural production - Food production and photosynthesis: Go online
Questions

Q1: Complete the table using the words listed.

Raw materials for photosynthesis	Essential requirements	Products of photosynthesis

Words: carbon dioxide, chlorophyll, light, oxygen, sugar, water.

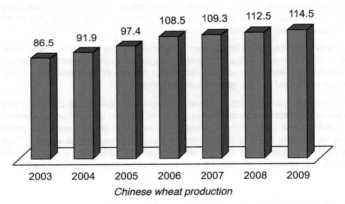

Chinese wheat production

Note: a bushel is technically a unit of capacity and so the weight of grain in a bushel depends on the type of grain. For wheat, one bushel is about 27 kilograms.

The increase in wheat yield in China is due to spread of technologies including modern irrigation projects, pesticides which protect crops from pests, synthetic (man-made) nitrogen fertilisers and improved crop varieties or cultivars with higher yields of wheat.

Q2: Calculate the percentage increase in wheat production in China from 2003 to 2009 to one decimal point.

....

Q3: Predict the level of wheat production in 2015 in million bushels.

....

Q4: Give two reasons why this might not be reached.

1.3 Agricultural production - Food production and trophic levels

Livestock produce less food per unit area than plant crops due to loss of energy between **trophic levels**.

	Trophic levels
Trophic level 4	Second level carnivores: eat first level carnivore
	↑
Trophic level 3	First level carnivores: eat herbivores
	↑
Trophic level 2	Herbivores: eat plants
	↑
Trophic level 1	Plants: produce energy from the sun and nutrients

Trophic levels

A food chain represents a succession of organisms that eat another organism and are, in turn, eaten themselves. The trophic level of an organism is the position it occupies in a food chain. The number of steps an organism is from the start of the chain is a measure of its trophic level.

Food chains start at trophic level 1 with primary **producers** such as plants, move to primary consumers (**herbivores**) at level 2, secondary consumers at level 3 and typically finish with tertiary consumers or apex **predators** at level 4. Secondary and tertiary consumers which feed on meat can be described as **carnivores**.

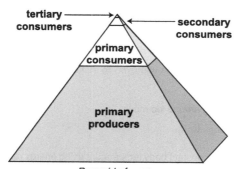

Pyramid of energy

The efficiency with which energy or **biomass** is transferred from one trophic level to the next is called the ecological efficiency. Consumers at each level convert on average only about 10 percent of the chemical energy in their food to their own organic tissue. For this reason, food chains rarely extend for more than 5 or 6 levels.

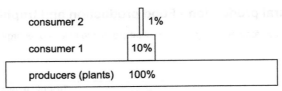

Pyramid of energy

The amount of available energy available in a food chain decreases from one stage to the next. Some of the available energy goes into growth and the production of offspring. This energy becomes available to the next stage, but most of the available energy is used up in other ways:

- energy released by respiration is used for movement and other life processes, and is eventually lost as heat to the surroundings;
- energy is lost in waste materials, such as faeces.

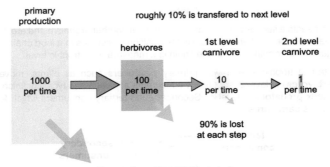

Energy loss along a food chain

The efficiency of food production can be improved by reducing the amount of energy lost to the surroundings. This can be done by:

- preventing animals moving around too much;
- keeping their surroundings warm.

Mammals and birds maintain a constant body temperature using energy released by respiration. As a result, their energy losses are high. Keeping pigs and chickens in warm sheds with little space to move around allows more efficient food production. But this raises moral concerns about the lives of such animals. In reality, a balance must be reached between the needs of farmers and consumers and the welfare of the animals.

Agricultural production - Trophic levels: Questions

Q5: From the information given in the diagram, how much energy in kJ is used in respiration?

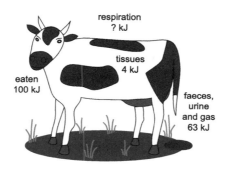

....

Q6: If only 4 kJ of the original energy available to the bullock is available to the next stage, which might be humans, what is the percentage efficiency of this energy transfer?

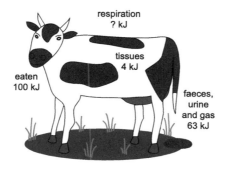

....

Q7: Why is energy transfer between trophic levels inefficient?

....

Q8: What happens to the energy that is not passed on at each stage of the food chain?

....

Q9: How would changing to a plant-based diet help reduce world hunger?

1.4 Agricultural production - Efficient food production

Food production is more efficient if the food chain is short. There are fewer **trophic levels** at which energy can be lost and therefore a higher percentage of energy is available to humans.

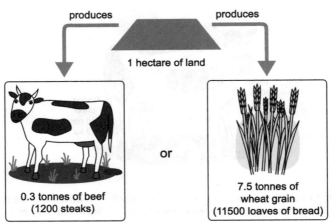

Efficient food production

On average about 10 percent of net energy production at one trophic level is passed on to the next level. The nutritional quality of material that is consumed also influences how efficiently energy is transferred, because consumers can convert high-quality food sources into new living tissue more efficiently than low-quality food sources.

One of the consequences of loss of energy at each trophic level is that shorter food chains are more efficient than longer ones, as more energy is available to the final consumer.

Therefore the food chain:

wheat grain → human

has two trophic levels and is more efficient than the food chain below which has three trophic levels:

grass → cow → human

and therefore passes on less energy to humans.

From the earliest times, humans, who were unable to consume grass, learnt how to put it to good use by becoming herdsmen, long before they became farmers. Keeping **livestock** in fact enabled them to make the best use of spontaneous plant growth of this kind, and still remains the best way of turning primary production to good account in areas where, because of latitude or altitude, low temperatures or the limited amount of daylight do not permit farming.

Livestock production may be possible in managed and wild habitats unsuitable for cultivation of crops.

1.5 Learning points

Summary

- Food security is the ability of human populations to access food of sufficient quality and quantity.

- As a result of increase in human population and concern for food security, there is a continuing demand for increased food production.

- Food production should be sustainable.

- Food production should not degrade the natural resources on which agriculture depends.

- All food production is dependent ultimately upon photosynthesis.

- Most human food comes from a small number of plant crops.

- Due to limited area to grow crops, increased food production will depend on factors that control plant growth, for example the use of fertilisers to increase crop yield.

- Factors that increase food production are the breeding of higher yielding cultivars and protecting crops from pests, disease and competition.

- Livestock produce less food per area than plant crops due to loss of energy between trophic levels.

- Livestock production may be possible in managed and wild habitats unsuitable for the cultivation of crops.

1.6 Extension materials

The material in this section is not examinable. It includes information which will widen your appreciation of this section of work.

Extension materials: Measures against falling fish stocks

The number of cod in the North Sea has decreased dramatically in the past 20 years. In 1980 roughly 300,000 tons of cod were caught by fishermen in the North Sea, by 2001 this had fallen to 50,000 tons, and it is continuing to fall. There are huge problems for ensuring the survival of the stock.

The seas have been overfished and the number of cod has fallen dramatically. The quota of fish that can be removed has been drastically reduced and areas of the North Sea have been closed to fishermen to give the cod time to reproduce and replenish themselves.

The European Union (EU) has stated that a 40,000 square mile area of the North Sea, almost a fifth of its entire area, will be off limits for a couple of months for a few years to cod, haddock and whiting fishermen during the crucial spawning period for the fish, as part of a desperate attempt to ensure that the cod stock is not wiped out. This will put many trawlermen out of work, but it has been argued that it is essential if there are to be any cod left in the North Sea.

Cod stocks have fallen to their lowest levels in the last hundred years and quotas for the white fish were cut by 40% by EU ministers. An area from the north of Scotland to the east of England will be closed to trawlermen who take cod, haddock and whiting in the same nets. The North Sea ban will last from mid-February until the end of April. This is the crucial spawning period for cod. Some fishermen will be allowed into the so-called "controlled zone" but these will be on the lookout for species which swim at higher level such as mackerel and the policy will be rigorously policed with on-board observers making sure that they do not catch any cod.

Norway, which manages North Sea fishing along with the EU, has already agreed to include other emergency measures such as forcing fishermen to apply for special permits and reporting what they catch in greater detail.

1.7 End of topic test

End of Topic 1 test   Go online

Q10: The following flow chart shows the energy flow in a field of potatoes during one year.

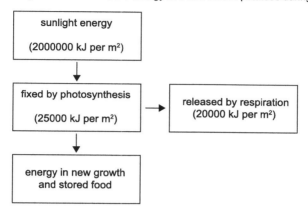

What percentage of the available sunlight energy would be present in new growth and stored food in the potato crop?

a) 2.25
b) 1.25
c) 1.00
d) 0.25

....

Q11: Organisms that obtain their energy directly from photosynthesis are known as:

a) herbivores.
b) producers.
c) first level carnivores.
d) second level carnivores.

....

Q12: Which of the following characteristics would a cultivar not be selected for and continually cultivated?

a) Increased yield of grain
b) Increased fruit production
c) Increased susceptibility to disease
d) Increased rate of growth

....

Q13: The following diagram represents a food chain consisting of four trophic levels A, B, C and D.

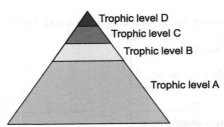

Based on the diagram above, the greatest amount of energy (and biomass) in a healthy food chain will be found in:

a) trophic level A: producers.
b) trophic level B: primary consumers.
c) trophic level C: secondary consumers.
d) trophic level D: 3rd (tertiary) consumers.

....

Q14: A woodland ecosystem receives about 1,000,000 kJ m^{-2} year^{-1} of solar energy. Of this, energy 96% is not used in photosynthesis. Which of the following shows the amount of energy captured by the producers in this ecosystem?

a) 400 kJ m^{-2}year^{-1}
b) 4,000 kJ m^{-2}year^{-1}
c) 40,000 kJ m^{-2}year^{-1}
d) 400,000 kJ m^{-2}year^{-1}

....

Q15: The following diagram shows a food chain consisting of four trophic levels.

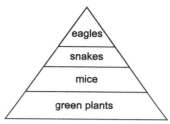

If 1,000 kJ of energy enters at the green plant trophic level, how much energy is available for use by the eagle?

a) 100 kJ
b) 10 kJ
c) 1 kJ
d) 0.1 kJ

....

Q16: Upon which process does all food production ultimately depend?

....

Q17: Why does the human population need to increase its rate of food production?

Unit 3 Topic 2

Plant growth and productivity

Contents

2.1	Photosynthesis and energy capture	263
2.2	Photosynthetic pigments	264
2.3	The spectrum of light	265
2.4	Absorption spectrum	266
2.5	Action spectrum	267
2.6	First stage of photosynthesis: The light-dependent stage	269
2.7	The second stage of photosynthesis: The carbon fixation stage	271
2.8	Learning points	273
2.9	Extended response question	274
2.10	End of topic test	275

Prerequisites

You should already know that:

- photosynthesis is a series of enzyme-controlled reactions, in a two-stage process;
- during the light reactions, light energy from the sun is trapped by chlorophyll in the chloroplasts and is converted into chemical energy in the form of ATP;
- water is split to produce hydrogen and oxygen (excess oxygen diffuses from the cell);
- during carbon fixation, hydrogen and ATP produced by the light reaction is used with carbon dioxide to produce sugar;
- the chemical energy in sugar is available for respiration or can be converted into plant products such as starch and cellulose.

Learning objective

By the end of this topic, you should be able to:

- describe the fates of light as it strikes a leaf;
- give two reasons plants absorb light energy;
- give the meaning of the term absorption spectrum;
- state that each pigment absorbs a different range of wavelengths of light;
- describe the role of carotenoids;
- give the meaning of the term action spectrum;
- compare the absorption spectrum of chlorophyll a and b and carotenoids to action spectra for photosynthesis;
- describe how light energy is used to generate ATP;
- describe photolysis;
- describe the carbon fixation stage;
- describe the fates of glucose produced by photosynthesis.

2.1 Photosynthesis and energy capture

Photosynthesis converts light energy into chemical energy stored in carbohydrates (sugars) and other organic compounds. This process consists of a series of chemical reactions that require carbon dioxide (CO_2) and water (H_2O). Light energy from the sun drives the reaction. The light energy is trapped by photosynthetic pigments such as **chlorophyll** and stored as chemical energy in the carbohydrates produced. Oxygen (O_2) is a by-product of photosynthesis and is released into the atmosphere. The following equation summarises photosynthesis:

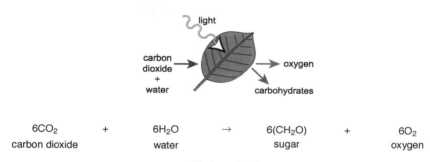

$$6CO_2 + 6H_2O \rightarrow 6(CH_2O) + 6O_2$$

carbon dioxide + water → sugar + oxygen

Photosynthesis

Only particular wavelengths of light that strike a leaf are **absorbed** by photosynthetic pigments but very little of the energy is actually converted into useful chemical energy. The rest of the light striking the leaf is either **reflected** off the leaf surface or is **transmitted** through the leaf. These processes are sometimes called the fates of light striking a leaf.

The fate of light striking a leaf

- Absorbed: Light is taken into the leaf (5%).
- Reflected: Light is bounced back from the leaf surface (85%).
- Transmitted: Light passes through the leaf (10%).

Of the light which is absorbed only a very small part is used for photosynthesis. The rest is lost, or radiated as heat from the leaf. Light is absorbed into organelles called **chloroplasts** that are found in the palisade layer of the leaf. Chloroplasts contain **chlorophyll**, a pigment that is essential for photosynthesis.

2.2 Photosynthetic pigments

The photosynthetic pigments absorb light energy and convert it into chemical energy. Chlorophyll a and b are the main photosynthetic pigments, they absorb mainly blue and red light wavelengths. The carotenoids are accessory pigments that absorb other wavelengths of light. The carotenoids extend the range of wavelengths absorbed and pass this energy onto **chlorophyll**.

Photosynthetic pigments: Thin layer chromatography   Go online

This activity illustrates the separation of photosynthetic pigments by thin layer chromatography.

In (A), the sample has been spotted onto the plate. In (B), the solvent has moved to the top of the plate and the photosynthetic pigments have been separated.

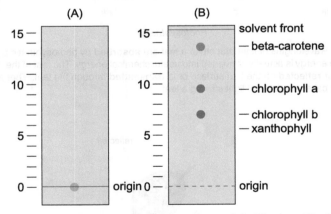

Each of the four photosynthetic pigments has a characteristic Rf value. The Rf value is calculated by dividing the distance moved by the pigment (at the front or leading edge) by the distance moved by the solvent.

In this experiment, the solvent has moved a distance of 15.3 units. The pigment beta-carotene (a carotenoid) has moved a distance of 14 units, so the Rf value for beta-carotene is $\frac{14}{15.3}$ or 0.92.

Q1: What is the R_f value for xanthophyll (a carotenoid)?

a) 0.39
b) 0.49
c) 0.59
d) 0.69

....

Q2: What is the R_f value for chlorophyll b?

a) 0.28
b) 0.38
c) 0.48
d) 0.58

....

Q3: What is the R_f value for chlorophyll a?

a) 0.34
b) 0.44
c) 0.54
d) 0.64

2.3 The spectrum of light

The spectrum of light can be seen if a beam of light is shone through a glass prism onto a screen. The spectrum is a rainbow of colours of different wavelengths.

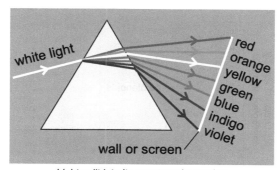

Light split into its spectrum by a prism

Colours of the spectrum of light: red, orange, yellow, green, blue, indigo and violet.

The spectrum of light: The colour spectrum of visible light   Go online

This activity shows what happens when visible light passes through a prism. It is the blue and red parts of visible light that are directly absorbed by chlorophyll.

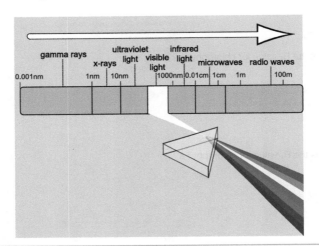

2.4 Absorption spectrum

The absorption spectrum shows the extent to which different colours of light are absorbed by a pigment. This can be shown as a graph.

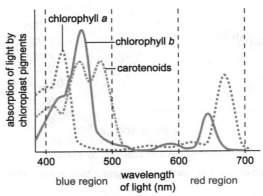

The absorption spectra of chlorophyll a, b and carotenoids

The graph shows that there is limited absorption between wavelengths 500-650 nm which is the green region of the spectrum with most light reflected or transmitted. This is why most plants are green in colour.

Chlorophyll a and b absorb mainly in the blue and red regions of the spectrum. The carotenoids extend the range of wavelengths available for photosynthesis and pass this energy onto **chlorophyll**.

2.5 Action spectrum

The action spectrum shows the rate of photosynthesis carried out in different wavelengths of light. Again, this can be shown as a graph.

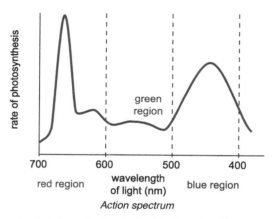

Action spectrum

The graph shows the rate of photosynthesis is highest in the red and blue regions of the spectrum.

If the action spectrum for photosynthesis is placed on top of the absorption spectrum for photosynthetic pigments, you can see that the two are very closely related. This shows that the pigments are involved in absorption of light for photosynthesis.

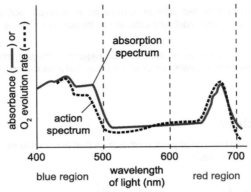

Absorption spectrum and action spectrum combined

Action spectrum: An experiment to determine an action spectrum for photosynthesis   Go online

The following experiment was carried out to investigate which parts of the visible light spectrum are used in photosynthesis.

- A sample of the aquatic plant Elodea, with the cut edge of the stem uppermost, was added to a beaker of water. Water was added to ensure the plant was fully covered.
- A lamp was placed 20 cm from the beaker to provide a light source.
- The plant was allowed to photosynthesise without treatment for 10 minutes.
- Coloured filters (blue, green, yellow and red) were placed at separate times between the lamp and the plant. The number of oxygen bubbles produced in a 2 minute period was counted.

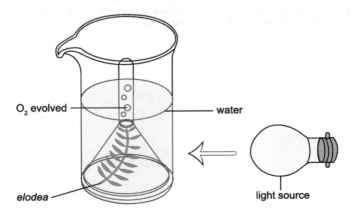

The following table shows the results of the experiment.

Colour of filter	Number of oxygen bubbles produced
Blue	20
Green	5
Yellow	9
Red	16

Q4: Which wavelengths of light are used for photosynthesis? Plot the results of the experiment on a graph with the number of oxygen bubbles produced on the y-axis and the colour of filter (the wavelength of light) on the x-axis.

....

Q5: What conclusion can you draw from these results?

....

Q6: Why is it necessary to allow the plant to photosynthesise without treatment before carrying out the experiment?

2.6 First stage of photosynthesis: The light-dependent stage

Photosynthesis is a series of enzyme-controlled reactions that occurs in the chloroplasts of plants.

- In photosynthesis, organic molecules such as carbohydrates and amino acids are **synthesised** by the reduction of carbon dioxide.
- The energy to drive the reactions comes from light.
- Light energy is **absorbed** into the pigments contained in plant cell **chloroplasts**.

The first stage of photosynthesis is dependent on light and is often referred to as the **light reaction**. If a pigment molecule absorbs light energy, an electron in the molecule becomes excited i.e. the electron's energy level is raised to become a high-energy electron. These high-energy electrons can then be transferred through the electron transport chain to bring about production of **ATP** by the enzyme ATP synthase.

The energy is also used for the photolysis of water. Water is split into oxygen which is evolved as a by-product of the reaction, and hydrogen which is transferred to the coenzyme **NADP** and combined to produce NADPH. The ATP and NADPH produced in the light reaction of photosynthesis are used in the next stage of photosynthesis referred to as the carbon fixation stage (**Calvin Cycle**).

The following activity summarises the light-dependent stage of photosynthesis.

The light-dependent stage

Q7: Complete the labelling of the diagram using the terms listed to the right.

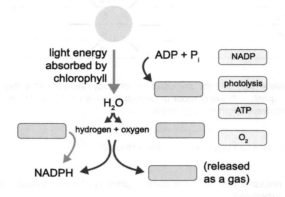

Q8: Complete the labelling of the diagram using the words and phrases listed underneath.

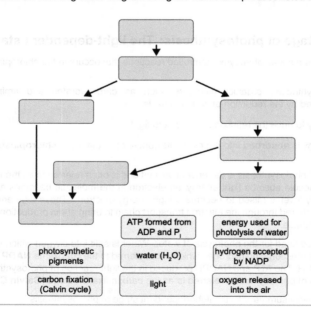

2.7 The second stage of photosynthesis: The carbon fixation stage

The second stage of photosynthesis is known as the carbon fixation stage or the **Calvin Cycle**. This stage does not require light. In this reaction carbon dioxide is converted into sugars using **ATP** and **NADPH** from the **light reaction**.

In the carbon fixation stage, the enzyme Ribulose-1,5-bisphosphate carboxylase oxygenase, commonly known by the shorter name **RuBisCO**, fixes carbon dioxide (CO_2) from the atmosphere. RuBisCO fixes carbon dioxide by attaching it to RibuloseBisPhosphate (**RuBP**) to form 3-phosphoglycerate. The 3-phosphoglycerate produced is phosphorylated by ATP and combined with hydrogen from **NADPH** to form the stable compound glyceraldehyde-3-phosphate (**G-3-P**). G-3-P is used to regenerate RuBP and for the synthesis of glucose. Glucose may be used as a respiratory substrate, synthesised into starch or cellulose or passed to other biosynthetic pathways. These biosynthetic pathways can lead to the formation of a variety of **metabolites** such as DNA, protein and fat.

The second stage of photosynthesis: The carbon fixation stage   Go online

Q9: Complete the carbon fixation diagram using the terms listed underneath.

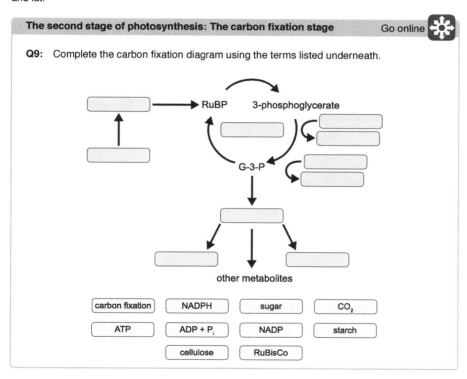

The second stage of photosynthesis: Questions

Go online

The following diagram summarises the process of photosynthesis in a chloroplast of a leaf.

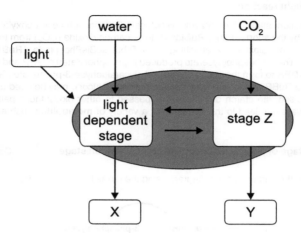

Use the diagram to answer the following questions.

Q10: Name the molecules X and Y.

....

Q11: In the light-dependent stage water is split up into its constituent components using light energy. What term is used to describe this splitting of water?

....

Q12: Name stage Z.

....

Q13: Molecule Y can be converted into a cell wall component. Name the cell wall component.

....

Q14: Name two molecules formed in the light-dependent stage which are required for stage Z.

2.8 Learning points

Summary

- Light striking a leaf is transmitted, reflected or absorbed.
- Chlorophylls a and b are the major pigments involved in absorption of light.
- The wavelengths of light that are absorbed by a photosynthetic pigment are called its absorption spectrum.
- The absorption spectrum of chlorophyll is closely related to the rate of photosynthesis.
- Absorption of light by chlorophyll occurs mainly in the blue and red regions of the spectrum.
- Carotenoids extend the range of wavelengths absorbed and pass the energy to chlorophyll for photosynthesis.
- The wavelengths of light actually used by a pigment in photosynthesis are called its action spectrum.
- Photosynthesis is a series of enzyme-controlled reactions that synthesise carbohydrates from carbon dioxide and water.
- Photosynthesis occurs in two stages: the light dependent stage and the carbon fixation stage.
- In the light dependent stage, light energy is used to regenerate ATP, absorbed light energy excites electrons in the pigment molecule, these high-energy electrons are transported through electron transport chain and generate ATP by ATP synthase.
- Light energy is also used to split a water molecule into hydrogen and oxygen, a process called the photolysis of water.
- Oxygen is evolved from the leaf as a by-product.
- The hydrogen is transferred to the carbon fixation stage by the hydrogen acceptor NADP that becomes reduced to form NADPH.
- The ATP from the light dependent stage is transferred to the carbon fixation stage.
- The enzyme RuBisCO fixes carbon dioxide from the atmosphere by attaching it to RuBP.
- The 3-phosphoglycerate produced is phosphorylated by ATP and combined with the hydrogen from NADPH to form G-3-P.
- G-3-P is used to produce sugars such as glucose which may be synthesised into starch, cellulose or other metabolites.
- G-3-P is also used to regenerate RuBP to continue the cycle.
- Major biological molecules in plants such as proteins, fats, carbohydrates and nucleic acids are derived from the photosynthetic process.

2.9 Extended response question

The activity which follows presents an extended response question similar to the style that you will encounter in the examination.

You should have a good understanding of photosynthesis before attempting the question.

You should give your completed answer to your teacher or tutor for marking, or try to mark it yourself using the suggested marking scheme.

Extended response question: Plant growth and productivity

A) Discuss the role of light and photosynthetic pigments in photosynthesis. *(8 marks)*

B) Give an account of the light-dependent stage of photosynthesis and the carbon fixation stage. *(8 marks)*

2.10 End of topic test

End of Topic 2 test Go online

Q15: The action spectrum of photosynthesis is a measure of the ability of plants to:

a) absorb all wavelengths of light.
b) absorb light of different intensities.
c) use light to build up foods.
d) use light of different wavelengths for photosynthesis.

....

Q16: The following diagram contains information about light striking a leaf.

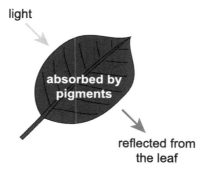

1. Apart from being absorbed or reflected, what can happen to light which strikes a leaf?
2. Pigments that absorb light are found within leaf cells. State the location of these pigments within leaf cells.

....

Q17: The following diagram shows part of the light dependent stage of photosynthesis.

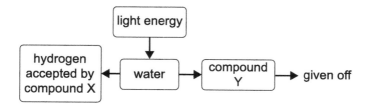

1. Name this part of the light dependent stage.
2. Name compound X.
3. Name compound Y.

Q18: As well as chlorophyll, plants have other photosynthetic pigments. State the benefit to the plant of having these other pigments.

Q19: The following diagram is a simplified version of the carbon fixation stage.

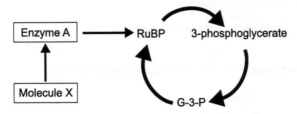

1. Enzyme A fixes molecule X by attaching it to RuBP. Name molecule X.
2. Name Enzyme A.

Q20: Hydrogen is formed in the light stage of photosynthesis and is required for the carbon fixation stage. Name ONE other substance which is produced in the light stage and is also required for the carbon fixation stage.

Q21: Which of the following changes in concentration of the chemicals RuBP and G-3-P would occur if an illuminated green plant cell's source of carbon dioxide were removed?

	RuBP	G-3-P
A	Increase	Increase
B	Decrease	Decrease
C	Increase	Decrease
D	Decrease	Increase

Q22: Complete the table about the first stage in the chemistry of photosynthesis by matching the terms to the descriptions.

Term	Description
	Product of the photolysis of water which is required for aerobic respiration.
	Compound which accepts hydrogen during the photolysis of water.
	Raw material which becomes split into oxygen and hydrogen during photolysis of water.
	Components of a high-energy compound.
	Breakdown of water during the light- dependent stage of photosynthesis.
	Product of photolysis of water which becomes attached to NADP.
	Green pigment which traps light energy.
	First stage in photosynthesis in which light energy is converted to chemical energy.

Terms: ADP + P_i, Chlorophyll, Hydrogen, Light dependent reaction, NADP, Oxygen, Photolysis, Water.

. . . .

Q23: Complete the table about the carbon fixation stage in the chemistry of photosynthesis by matching the terms to the descriptions.

Term	Description
	Structure found in a leaf where photosynthesis takes place.
	Hydrogen acceptor needed for the fixation of carbon in carbohydrates.
	Carbon compound which acts as a carbon dioxide acceptor.
	First stable compound formed in the carbon fixation stage (Calvin Cycle) after carbon dioxide combines with its acceptor molecule.
	Raw material which supplies carbon atoms to be fixed into carbohydrates.
	High-energy compound used to phosphorylate the intermediate compound in carbon fixation (Calvin Cycle).
	Second stage in photosynthesis which is also known as the carbon fixation stage (Calvin Cycle).
	The enzyme which fixes carbon dioxide by attaching it to RuBP.

Terms: ATP, Carbon dioxide, Carbon fixation, Chloroplast, G-3-P, NADPH, RuBisCO, RuBP.

....

Q24: In an investigation, the rate of photosynthesis by nettle leaf discs was measured at different light intensities. The results are shown in the following table.

Light intensity (Kilolux)	Rate of photosynthesis by nettle leaf discs (Units)
10	4
20	28
30	60
40	90
50	92
60	92

1. Plot a line graph to show the rate of photosynthesis by nettle leaf discs at different light intensities. Use appropriate scales to fill most of the graph paper. *(2 marks)*
2. From the table, predict how the rate of photosynthesis at light intensities of 50 kilolux could be affected by an increase in carbon dioxide concentrations. Justify your answer.

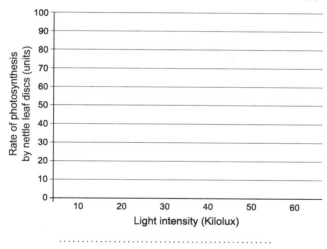

Q25: An investigation was carried out into the effect of carbon dioxide concentration on yield. Tomato plants were cultivated in glasshouses, where it was possible to control the concentration of carbon dioxide in the atmosphere.

The carbon dioxide concentrations ranged from 50 to 1200 parts per million (ppm) per volume. The yield of tomatoes was measured in kg per m^2. The temperature and light intensity conditions were constant for all concentrations of carbon dioxide.

The results are shown in the following graph.

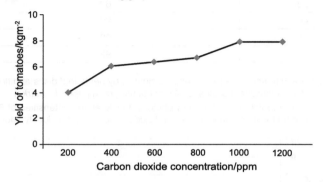

1. Describe the effect of increasing the carbon dioxide concentration on the yield of tomatoes. (*2 marks*)
2. The normal concentration of carbon dioxide in the atmosphere is 300 ppm. From the graph, determine the yield of tomatoes when the concentration in the glasshouse was 300 ppm.
3. Calculate the percentage change in yield that would be expected if the tomatoes were grown in an atmosphere where the carbon dioxide concentration was increased to 1000 ppm compared with the yield at 200 ppm.

....

Q26: Explain why carbon dioxide concentration affects the yield of tomatoes.

....

Q27: A further experiment was carried out to investigate the effect of temperature on yield. In this experiment the carbon dioxide concentration was 300 ppm and the light intensity remained constant.

The results are shown in the following graph.

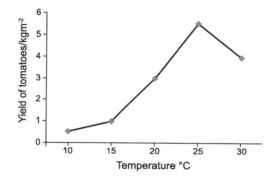

1. Compare the yield of tomatoes at 15°C with that at 25°C and suggest an explanation for the difference in yield. (*2 marks*)
2. Calculate the percentage increase in yield of tomatoes from 15°C to 25°C.
3. Name one factor, other than light, temperature and carbon dioxide concentration, which could affect the yield of tomatoes grown in a glasshouse. (*1 mark*)
4. Suggest an explanation for the shape of the graph between 25°C and 30°C.

Unit 3 Topic 3

Plant and animal breeding

Contents

3.1 Introduction 285
3.2 Field trials 286
3.3 Inbreeding 287
3.4 Cross breeding and F1 hybrids 288
3.5 F1 Hybrids 289
3.6 Genetic technology 289
3.7 Learning points 291
3.8 Extension materials 293
3.9 End of topic test 294

Prerequisites

You should already know that:

- an increasing human population requires an increased food yield;
- GM crops may be an alternative to mitigate the effects of intensive farming on the environment;
- alleles are the different forms of a gene;
- homozygous means having the same alleles for a particular gene;
- heterozygous means having different alleles for a particular gene.

Learning objective

By the end of this topic you should be able to:

- state that plant and animal breeding involves the manipulation of heredity;
- state that this manipulation will allow the development of new and improved organisms to provide sustainable food sources;
- describe the characteristics that breeders seek to develop in crops and stock;
- describe the process of artificial selection;
- explain the need for plant field trials;
- explain that trials are carried out in a range of environments to compare the performance of different cultivars or treatments;
- explain the factors which have to be taken into account when designing field trials including:
    - the selection of treatments to ensure fair comparison;
    - the number of replicates to take account of the variability within the sample;
    - the randomisation of treatments to eliminate bias when measuring treatment effects;
- describe the process of inbreeding;
- understand that inbreeding depression is the accumulation of recessive, deleterious homozygous alleles;
- explain how new alleles can be introduced to plant and animal lines;
- describe how, in animals, individuals from different breeds may produce a new crossbreed population with improved characteristics;
- state that the F2 generation will have a wide variety of genotypes;
- state that, in plants, F1 hybrids, produced by the crossing of two different inbred lines, creates a relatively uniform heterozygous crop;
- state that F1 hybrids often have increased vigour and yield;
- describe how as a result of genome sequencing, organisms with desirable genes can be identified and then used in breeding programmes;
- explain that genetic transformation techniques allow a single gene to be inserted into a genome;
- describe how this reprogrammed genome can be used in breeding experiments.

3.1 Introduction

Rice, wheat, maize, barley, sorghum, millet and sugar cane are grasses that have been adapted for human use. Humans have bred these grasses for plumper seeds, taller stems, earlier ripening and resistance to drought, rain, insects or disease. The area of land given over to crops such as these continues to increase. Together grasses supply around 15 percent of carbohydrates and more than 50 percent of protein to the human population.

All the plants in the figure below are derived from one wild species, the wild cabbage, *Brassica oleracea*. Humans have taken this wild plant and, over the centuries, shaped it into these different kinds of foods. This form of selection in which humans have improved characteristics in organisms is known as **artificial selection**.

Artificial selection has generated diversity in both plants and animals. In **agriculture**, superior strains of corn, wheat, and soybeans have resulted from careful breeding.

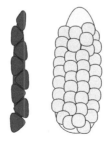

Teosinte (left) and its modern descendent, corn (right), a product of artificial selection.

Plant and animal breeding involves the manipulation of heredity to develop new and improved organisms to provide sustainable food sources. Breeders seek to develop crops and stock with higher yield, higher nutritional values and resistance to pests and diseases. Other physical characteristics suited to rearing and harvesting have also been developed, as well as those characteristics which enable crops and stock to thrive in particular environmental conditions.

Wheat is one example of a crop which has undergone artificial selection for centuries. By the mid 1900s farmers had managed to produce high yielding wheat plants but they were so tall that the heavy seed heads caused them to fall over in the wind, making the seeds fall to the ground and rot. By crossing a tall high yielding variety of wheat with a dwarf variety, farmers were able to produce a short plant with a high yield which could withstand windy conditions.

3.2 Field trials

The aim of plant breeding is to produce **cultivars** that will have good yield in the growth conditions typical for that crop. Final crop growth is a result of both genetic and environmental factors; a new plant variety may grow well in a laboratory environment however its performance will have to be evaluated in field trials. Plant field trials are carried out in a range of environments to compare the performance of different cultivars or treatments. Field trials can also be used to evaluate genetically modified (GM) crops.

Field trials have to be carefully and scientifically monitored to ensure accurate results are obtained and there are no adverse effects on the environment. Field trials are carried out in an area of land which is divided into plots. Each plot is given a different treatment for example varying herbicide concentration or used to grow a different variety of crop.

Plant field trials

In designing plant field trials account has to be taken of the selection of treatments, the number of replicates and the randomisation of treatments.

The selection of treatments must be considered to ensure fair comparisons. For example several plots may be given different quantities of herbicide to ensure a variety of treatments can be compared. There must be an adequate number of replicates to take account of the variability within the sample; in other words several replicates of each plot must be performed. The treatments must also be randomised to eliminate bias when measuring treatment effects. Randomisation means ensuring that replicate plots are not always in the same orderly sequence.

Field trials may show that a new variety of plant is not only superior to existing ones but is capable of growing in conditions where existing varieties would fail. For example maize for animal fodder can now be grown in northern parts of Scotland and most of the wheat for bread making in the UK is now grown in the UK rather than being imported from North America as a result of the development of new varieties.

Field trials: Question   Go online

Q1: Complete the following table by matching the design features of field trials to the reasons for carrying out procedures.

Design feature	Reason for carrying out this procedure
	To eliminate bias when measuring treatment effects.
	To take account of the variability within a sample.
	To ensure fair comparison.

Design features: Selection of treatments; Number of replicates; Randomisation of treatment.

3.3 Inbreeding

Outbreeding is the mating or breeding of unrelated individuals. Outbreeding often prevents the expression of deleterious (harmful) traits because the recessive **allele** controlling the trait is masked by a dominant allele.

Inbreeding is the mating or breeding of closely related individuals. When inbreeding is performed for long periods of time, there is a loss of **heterozygosity**. This means plants or animals become **homozygous** for the trait being selected. In inbreeding selected plants or animals are bred for several generations until the population breeds true to the desired type due to the elimination of heterozygotes. This means all the offspring are homozygous for the desired trait i.e. disease resistance.

Although inbreeding can select for desirable characteristics the accumulation of other homozygous alleles can cause the expression of deleterious (harmful) recessive alleles. This is referred to as **inbreeding depression** and may result in reduced yield. This generally leads to a decreased fitness of a population.

The cross below shows what can happen to a variety of wheat when inbreeding occurs. There is a chance that a combination of two recessive deleterious (harmful) alleles (aa) will be produced.

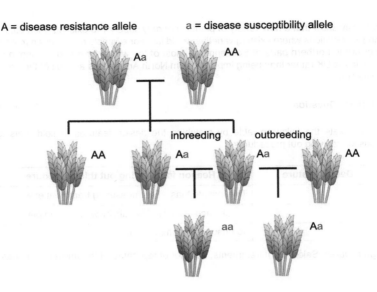

Some of the consequences of inbreeding depression include; reproductive failures, poor health, small litters, reduced immune system, high susceptibility to infections and shorter lives.

3.4 Cross breeding and F1 hybrids

Inbreeding cannot be carried out indefinitely; eventually deleterious **alleles** will accumulate and cause **inbreeding depression**. New alleles can be introduced to plant and animal lines by crossing a **cultivar** or breed with an individual with a different, desired **genotype**.

For example, different breeds of sheep show variation in their fertility, rate of meat production, disease resistance and wool quality. A breeder may choose to **crossbreed** a variety of sheep which has a good rate of meat production with a different breed which has a high lambing rate. The offspring of this mating are known as the F1 generation and they receive half their genetic information from one parent and half from the other. A series of back crosses are then performed to eliminate any unwanted genetic material from the new breed, whilst maintaining the desired characteristic.

A backcrossing is a crossing of a heterozygous organism showing a desirable characteristic with one of its homozygous parents showing the same desirable characteristic. The offspring showing the desired characteristics are selected to parent the next generation and another backcross is performed. This process is repeated until much of the unwanted genetic material is lost and only the desirable characteristics remain.

TOPIC 3. PLANT AND ANIMAL BREEDING

In plants, F1 hybrids, produced by the crossing of two different inbred lines, create a relatively uniform heterozygous crop. F1 hybrids often have increased vigour and yield. As an F2 population will have a wide variety of genotypes, a process of selection and backcrossing is required to maintain the new breed. Alternatively the two parent breeds can be maintained to produce more heterozygous F1 hybrids.

3.5 F1 Hybrids

Two parent plants from different inbred lines can be crossed to produce hybrid offspring. These offspring are heterozygous, usually show a uniform phenotype and are known as F1 hybrids. F1 hybrids often have increased vigour. This means they have a bigger yield and/or increased fertility or have other improvements compared to their parents. In the following example, the hybrid plant has a greater yield than either parent.

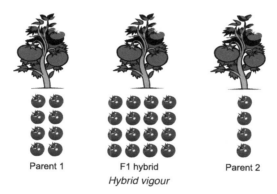

Parent 1   F1 hybrid   Parent 2
Hybrid vigour

F1 hybrids have a high degree of heterozygosity. This means when the F1 generation is self-crossed, the F2 generation will show a variety of genotypes. The F2 generation is described as genetically variable and is of little use for further production although it can provide a source of new varieties.

3.6 Genetic technology

As humans have made advances in technology, genetic sequencing and genetic transformation now play an important role in plant and animal breeding.

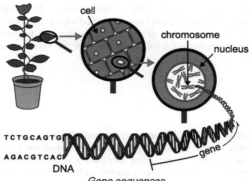
Gene sequences

Genome sequencing involves discovering the sequence of bases within an organism's genome. The information generated by this process can be used to identify organisms with specific desirable gene sequences. This organism can then be used in breeding programmes to produce offspring also showing the desirable characteristic.

Artificial selection can only draw on the natural variation within a species. Breeders sometimes want to combine a desirable characteristic from one species with another species. If genetic information from one species is to be transferred to another, techniques of genetic engineering are used.

Genetic transformation techniques allow a single gene to be inserted into a genome and this genome can then be used in breeding programmes. For example, insect resistant cotton uses a gene from a naturally occurring soil bacterium to provide it with built-in insect protection. The use of insect resistant cotton has reduced pesticide use over 80 percent in Australia.

Glyphosate

Glyphosate is a widely available herbicide which is extensively used by farmers and gardeners to control weeds. Genetically modified soybeans containing the glyphosate resistance gene for herbicide tolerance are now commercially available. Along with other agricultural crops such as soy, maize, alfalfa, canola and sorghum, these genetically modified plants have had the glyphosate gene inserted into their own genome. These cultivars greatly improve conventional farmers' ability to control weeds, since glyphosate could be sprayed on fields without damaging the crop. As of 2013, 90% of U.S. soybean fields were planted with glyphosate-resistant varieties.

Bt toxin

Bt toxin is produced by the bacterium *Bacillus thuringiensis* (Bt) in an inactive form (protoxin), which is transformed to its active form (delta-endotoxin) in the guts of certain insects. The active toxin binds to receptors in the gut, killing the insect. By means of genetic engineering, the genes for the active agent (Bt toxin) can be transferred from Bt bacteria to plants. There they produce the toxic agent inside the plant cells. In this way, biotechnology has been used to confer insect resistance to a number of economically important crops. Bt maize and Bt cotton are widely grown in several countries.

TOPIC 3. PLANT AND ANIMAL BREEDING

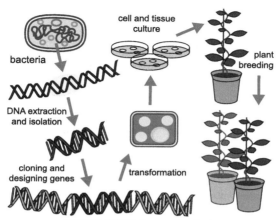

Transfer of Bt toxin gene to plants

3.7 Learning points

Summary

- Plant and animal breeding allows breeders to improve characteristics to help support sustainable food production.

- Breeders develop crops and animals with higher food yields, higher nutritional values, pest and disease resistance and ability to thrive in particular environmental conditions.

- Plant field trials are carried out in a range of environments to compare the performance of different cultivars or treatments and to evaluate GM crops.

- In designing field trials account has to be taken of the selection of treatments, the number of replicates and the randomisation of treatments.

- The selection of treatments to ensure valid comparisons, the number of replicates to take account of the variability within the sample, and the randomisation of treatments to eliminate bias when measuring treatment effects.

- In inbreeding, selected related plants or animals are bred for several generations until the population breeds true to the desired type due to the elimination of heterozygotes.

- A result of inbreeding can be an increase in the frequency of individuals who are homozygous for recessive deleterious alleles. These individuals will do less well at surviving to reproduce. This results in inbreeding depression.

- In animals, individuals from different breeds may produce a new crossbreed population with improved characteristics. The two parent breeds can be maintained to produce more crossbred animals showing the improved characteristic.

Summary continued

- In plants, F1 hybrids, produced by the crossing of two different inbred lines, create a relatively uniform heterozygous crop. F1 hybrids often have increased vigour and yield.

- In inbreeding animals and plants, F1 hybrids are not usually bred together as the F2 produced shows too much variation.

- New alleles can be introduced to plant and animal lines by crossing a cultivar or breed with an individual with a different, desired genotype.

- Plants with increased vigour may have increased disease resistance or increased growth rate.

- As a result of genome sequencing, organisms with desirable genes can be identified and then used in breeding programmes.

- Single genes for desirable characteristics can be inserted into the genomes of crop plants, creating genetically modified plants with improved characteristics.

- Breeding programmes can involve crop plants that have been genetically modified using recombinant DNA technology.

- Recombinant DNA technology in plant breeding includes insertion of Bt toxin gene into plants for pest resistance and glyphosate resistance gene inserted for herbicide tolerance.

3.8 Extension materials

The material in this section is not examinable. It includes information which will widen your appreciation of this section of work.

Extension materials: The cheetah

The cheetah originated about 4,000,000 years ago, long before the other big cats. The oldest fossils place it in North America in what is now Texas, Nevada and Wyoming. Cheetahs were common throughout Asia, Africa, Europe and North America until the end of the last Ice Age, about 10,000 years ago, when massive climatic changes caused large numbers of mammals to disappear. About that time all cheetah in North America and Europe and most of those in Asia and Africa vanished.

Some experts think our present populations were derived from inbreeding by those very few surviving and closely related animals. This inbreeding "bottleneck", as theorised, led to the present state of cheetah genetics. All cheetahs are very genetically similar to each other and as a result, the 12,500 cheetahs that are alive now do not have enough genetic diversity to protect them from a micro-biological crisis such as a bad feline virus. If one cheetah dies from a virulent pathogen that comes along, there is a strong chance that a very high proportion of cheetahs could die from it also. This unusually low genetic variability in cheetahs is accompanied by a very low sperm count, reduced sperm motility, and deformed flagella on the sperm. These are the consequences of inbreeding depression in cheetahs.

3.9 End of topic test

End of Topic 3 test Go online

Q2: Give one example of a characteristic which breeders may attempt to improve in a plant or animal species.

....

Q3: In inbreeding, selected plants or animals are bred for several generations until the population breeds true to the desired type due to the elimination of _____.

Choose from:

- homozygotes;
- heterozygotes.

....

Q4: Pedigree dogs are produced by mating members of the same breed with one another. This often results in the production of offspring suffering conditions which affect their fitness. For example, bulldogs have problems with their breathing and Labradors are prone to arthritis.

This phenomenon is know as:

a) hybrid depression.
b) natural selection.
c) inbreeding depression.
d) selective breeding.

....

Q5: The bacterium *Bacillus thuringiensis* produces a substance called T-toxin that is poisonous to leaf-eating insects. The following information shows some of the procedures used by genetic engineers to insert the gene for the production of T-toxin into crop plants.

Procedure 1	chromosome extracted from the bacterial cells
	↓
Procedure 2	position of T-toxin gene located
	↓
Procedure 3	T-toxin gene cut out from bacterial chromosome
	↓
Procedure 4	T-toxin gene transferred into nucleus of host plant
	↓
Procedure 5	plant cells containing T-toxin gene grown into small plants

Explain why such genetically engineered crop plants would grow better than unmodified crop plants.

....

Q6: These crops were commercially successful for several years. However, they have since become susceptible to attack by some members of a particular insect species. Suggest a reason that would account for this observation.

....

Q7: In order to produce a supply of hybrids showing genetic uniformity, horticulturists often maintain two different true-breeding parental lines of a species of bedding plant. The hybrids cannot be used as the parents of the next generation because:

a) hybrids of annual plants always form sterile seeds.
b) the hybrids are heterozygous and therefore not true breeding.
c) a high mutation rate occurs amongst hybrid gametes.
d) hybrid vigour cannot be passed onto the next generation.

....

Q8: *Sward height* (the height of the vegetation) is a useful practical indicator of the availability of herbage (grass and other plants) to grazing animals. Sward surface height can be measured by placing a ruler vertical so that the lower edge just touches the ground. The height of the tallest leaf at that point is recorded. The following graph shows the influence of sward surface height on milk production in kg day^{-1} from spring-calving cows.

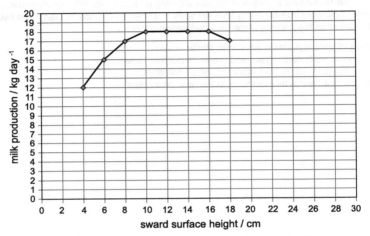

Describe the relationship between milk production and the sward surface height.

....

Q9: Using the information on the graph, suggest a suitable minimum sward height needed for grazing lactating dairy cows.

Unit 3 Topic 4

Crop protection

Contents

4.1 Introduction 299
4.2 Weeds, pests and diseases 300
4.3 Control of weeds, pests and diseases 301
4.4 Problems with pesticides 303
4.5 Biological control and integrated pest management 306
4.6 Learning points 308
4.7 End of topic test 310

Prerequisites

You should already know that:

- pesticides are used to kill unwanted pests on crops;
- pesticides which are sprayed onto crops can accumulate in the bodies of organisms over time;
- as pesticides are passed along food chains, toxicity increases and can reach fatal levels;
- biological control involves using natural predators to kill the pests;
- biological control may be an alternative to mitigate the effects of intensive farming on the environment.

> **Learning objective**
>
> By the end of this topic you should be able to:
>
> - explain why crop protection is important;
> - state that pests and weeds compete with crop organisms for resources, so yield is reduced;
> - describe the properties of annual weeds;
> - describe the properties of perennial weeds;
> - state that most of the pests of crop plants are invertebrate animals such as insects, nematode worms and molluscs;
> - state that plant diseases can be caused by fungi, bacteria or viruses, which are often carried by invertebrates;
> - state that pests and weeds can make infection by pathogens more likely;
> - state that chemical control of pests involves using herbicides, fungicides or insecticides;
> - understand that chemical control relies on use of chemical pesticides to kill pests;
> - describe the roles of herbicides, fungicides, pesticides and insecticides;
> - describe the differences between selective and systemic pesticides;
> - explain that protective applications of fungicide based on disease forecasts are often more effective than treating a diseased crop;
> - describe the potential problems associated with pesticides including:
>   - toxicity to animal species;
>   - persistence in the environment;
>   - accumulation in food chains;
>   - magnification in food chains;
>   - production of resistant populations;
> - understand that biological control involves using natural predators or parasites to kill the pests;
> - describe some of the risks associated with biological control;
> - explain that integrated pest management combines chemical, biological and cultural control.

TOPIC 4. CROP PROTECTION

4.1 Introduction

Monoculture is the agricultural practice of producing or growing one single crop over a wide area. It is widely used in modern industrial agriculture and its implementation has allowed for large harvests from minimal labour. Due to this practice in agriculture, crop protection is essential to ensure a sustained supply of good quality harvests. Crop protection is the branch of horticulture concerned with protecting crops from pests, weeds and disease.

Monoculture

Today many agricultural systems use monoculture, which involves expansion of land devoted to single crops and year-to-year production of same crop species on the same land. In these conditions **weed** competitors and pest and disease populations can multiply rapidly, reducing sustainability.

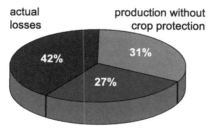

Crop production

The diagram above is an estimation of the contribution made by world-wide crop protection to the production of eight principal food and cash crops. As can be seen in the diagram, losses prevented by chemical crop protection are about 27% (weeds 16%, animal pests 7% and diseases 4%) of attainable production in the eight principal crops. Nevertheless, about 42% of the total production of these crops is lost to weeds (13%), animal pests (16%) and pathogens (13%), representing a market value of about 244 billion US$.

4.2 Weeds, pests and diseases

A **weed** in a general sense is a plant that is considered to be a nuisance, and is normally applied to unwanted plants in human-controlled settings, especially farm fields and gardens. Generally, a weed is a plant in an undesired place.

Annual weed plants grow, flower, set seed and die all within the space of one year. Due to the fact that they are short-lived, they are often vigorous and are a common problem in cultivated areas such as farmland, where constant tilling of the soil unearths seeds that germinate and grow quickly. The properties of annual plants that make them successful weeds include rapid growth, short life cycle, high seed output and long-term seed viability.

Perennial weed plants live for more than two years. Perennial weeds are able to compete with crop plants because they are already well established in the area the crop is being planted. Perennial weeds have storage organs which provide food when rates of photosynthesis are low. Some perennial weeds are also capable of vegetative reproduction; this means they have reproductive structures such as bulbs and tubers which new plants can grow from.

Tubers and a bulb

Most of the pests of crop plants are invertebrate animals such as insects, **nematode** worms, and **molluscs**. Insects cause damage to plant crops by feeding on them. Damage caused to leaves can reduce the rate of photosynthesis and therefore reduce crop yield. Nematode worms are found in the soil and attack plant's root systems. Snails and slugs are examples of molluscs; they cause damage to plants by eating them.

Plant diseases can be caused by fungi, bacteria or viruses. Invertebrates often act as vectors, facilitating the spread of diseases caused by microorganisms. Diseased plants have a reduced yield and a proportion of the crop may be unmarketable due to its appearance. Bacterial speck disease in tomato plants is caused by a bacterium called *Pseudomonas syringae*; this results in black lesions on the fruit, leaves and stems. Bacterial speck disease results in reduced growth due to a lower rate of photosynthesis and therefore reduced yield.

Bacterial speck disease symptoms

4.3 Control of weeds, pests and diseases

Weeds, pests and diseases can be controlled by cultural means. This form of control does not require chemicals but involves farmers adopting practices which make the environment unfavourable for the pest. One example of control of weeds, pests and diseases by cultural means is crop rotation. Crop rotation involves growing a different crop in the same area each successive year, this denies pests repeated access to their food source. A further example is ploughing, this buries crop residues that frequently harbour pests and diseases.

Other examples include:

- **polyculture**: the planting of several crops together in the same field. For instance, intercropping cowpeas with cassava reduces the abundance of whiteflies on the cassava;

- trap crops; plants grown amongst a main crop simply because they are more attractive to pests. For instance, Indian mustard is more attractive than cabbage for diamondback moths and leafrollers. In India, farmers grow one row of mustard between every five rows of cabbage allowing caterpillar populations build up on the mustard which can be treated by handsprayer with a high dosage of **insecticide**;

Diamondback moth: pupa (left) and larva stages

- **sanitation**; refers simply to the removal of crop residues and unharvestable plants that might harbour pest insects from outside the crop area;

- in some cases planting times can be delayed such that a crop is planted after a pest has emerged and died off. For instance, hessian fly populations are monitored in Georgia, USA, and farmers are advised when it is safe to plant their wheat crop.

Hessian fly

Cultural means of controlling weeds, pests and diseases aim to prevent their spread, often this is not enough to ensure a high crop yield and pesticides must be used. Pesticides provide farmers with a cost-effective way of improving the yield and quality of their crops. They also make harvesting more straightforward and maintain consistent yields from year to year.

Pesticides are chemicals which are used to kill pests such as insects, **nematodes** and **molluscs**. Pesticides include herbicides to kill weeds, fungicides to control fungal diseases, insecticides to kill insect pests, molluscicides to kill mollusc pests and nematicides to kill nematode pests.

Systemic insecticides, molluscicides and nematicides spread through the vascular system of plants. When a pest feeds on the plant they ingest the chemical and die. Systemic pesticides can be incorporated into the soil of crop plants. The chemical is absorbed by roots and translocated to leaves, stems, and flowers. An insect that feeds on a treated plant may acquire a lethal dose of insecticide or at least be deterred from further feeding.

Herbicides are chemicals used to kill weeds. Herbicides can be applied to control the growth of weeds which would otherwise grow amongst a crop, competing with it for water, nutrients and sunlight and reducing its yield.

Selective herbicides can be applied to crops which are established in a field; because they are selective they only kill the weeds. Many selective herbicides contain synthetic plant hormones which encourage the growth of plants which absorb them. Weeds often have broad leaves and take up large quantities of the chemical. This causes their growth to speed up; they use up their food stores and die. Crop plants such as cereals have narrow leaves therefore they do not take up as much of the chemical and are largely unaffected. Selective herbicides mimic natural plant hormones and therefore do not cause harm to the environment.

Broad leaved weed (left), narrow leaved crop plant (right)

Systemic herbicides kill all plant matter they come into contact with. They can be used to prepare a field before planting to clear it of all weeds. When systemic herbicides are sprayed on plants, they are absorbed and transported throughout the plant. This kills all parts of the plant including any reproductive structures under the soil. Systemic herbicides are biodegradable therefore they do not persist in the environment.

Fungicides are used to kill fungi which can cause disease in plants. Systemic fungicides are absorbed by crops and transported to all parts of the plant giving them protection from disease-causing fungi. Different fungal diseases require different environmental conditions. For example potato blight is caused by a fungus which requires specific humidity levels and temperatures to grow. By monitoring these conditions, farmers can give their crops protective applications of fungicide to prevent growth of the fungus. Protective applications of fungicide based on disease forecasts are often more effective than treating a diseased crop.

Potato blight: potato infected by the fungus Phytophthera

4.4 Problems with pesticides

Pesticides may be toxic to animal species. Chemical **insecticides** are generally intended for particular insect pests. Nevertheless, problems often arise because these chemicals are usually toxic to a broader range of organisms.

The potential for devastation is illustrated by the consequences of applying massive doses of dieldrin (an insecticide) to large areas of Illinois (USA) farmland from 1954 to 1958 in order to eradicate a grassland pest, the Japanese beetle *Popillia japonica*. Cattle and sheep on the farm were poisoned; 90% of the cats on the farms and a number of dogs were killed; and amongst the wildlife, 12 species of mammals and 19 species of birds suffered with meadow larks, robins, brown thrashers, starlings and ring-necked pheasants being virtually eliminated from the area.

Some pesticides may be persistent in the environment, this means they are not biodegradable and remain in the environment for long periods of time after their application. Pesticides can also accumulate within the body of an organism. Even if levels of the chemical in the environment are relatively low accumulation can occur if the organism absorbs the chemical at a faster rate than it is excreted (lost). This accumulation can mean that chemicals build up to toxic levels in an organism and cause poisoning.

Some pesticides are magnified along food chains. This means the concentration of the chemical increases as it moves from one trophic level to the next. The following diagram shows the process of biomagnification.

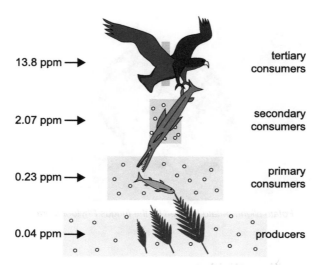

DDT is an insecticide which can be used to kill lice, fleas, greenfly and other insects. First used in Britain in 1939, it was important in controlling disease-carrying insects such as fleas and mosquitoes. It was also extensively used as a garden insecticide.

DDT is an example of a pesticide which is persistent; it also accumulates in the tissues of organisms and can be magnified along food chains.

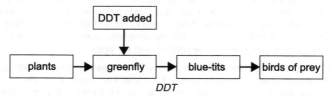

DDT is not biodegradable and therefore it persists in the environment for long periods of time. It is magnified along food chains and can accumulate in quantities large enough to cause thinning of egg shells, reducing successful reproduction rates. By the 1950s and 1960s, populations of birds of prey (such as sparrow hawks, peregrine falcons and eagles) were decreasing, and DDT was banned in Britain.

The use of pesticides may result in a population selection pressure producing a resistant population. The Colorado potato beetle has a legendary ability to develop resistance to a wide range of pesticides used for its control. The high beetle fecundity (birth) rate (on average, about 600 eggs per female) increases the probability that one of the numerous offspring mutates and develops resistance to the pesticide. With the very high fecundity, the pesticide resistance can spread rapidly through the population.

Colorado beetle

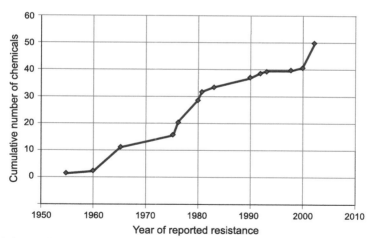

Cumulative number of insecticides to which resistance in the Colorado potato beetle has been reported

Problems with pesticides: Questions

The next two questions refer to the accompanying table which shows the concentration of a non-biodegradable pesticide residue in the tissues of the organisms in a marine food chain and the sea-water in their ecosystem.

Phytoplankton → krill → mackerel → tuna fish → dolphin

	Concentration of pesticide (ppm)
Water	0.00001
Phytoplankton	0.05
Krill (herbivore)	0.25
Mackerel (carnivorous fish)	2.0
Tuna fish	6.5
Dolphin	20

Q1: The concentration of pesticide increased by a factor of 400 times between:

a) water and tuna fish
b) krill and dolphin
c) phytoplankton and tuna fish
d) phytoplankton and dolphin

....

Q2: The concentration of pesticide in the tuna fish is greater than that in the phytoplankton by a factor of:

a) 5
b) 10
c) 130
d) 400

4.5 Biological control and integrated pest management

In biological control the control agent is a natural predator, parasite or pathogen of the pest. Examples of biological control include:

- using the caterpillar moth to kill cacti; the caterpillars feed on the cacti;
- using a parasitic wasp to kill whitefly; the wasp lays its eggs in the larvae of the whitefly;
- using a virus to kill rabbits; the myxoma virus kills rabbits;
- using ladybirds to kill aphids; ladybirds prey upon aphids;

- using bacteria to control blackfly caterpillars; the bacteria infects caterpillars and kills them using a poison.

Biological control (http://www.flickr.com/photos/ektogamat/2578779839/in/photostream/ by http://www.flickr.com/photos/ektogamat/, licensed under http://creativecommons.org/licenses/by/2.0/)

Advantages of biological control include its specificity, the predator/parasite only kills the pest. The predators will breed in the environment, so they do not need to be constantly reapplied and they do not cause harm to other organisms or accumulate in food chains. The predator/parasite is unlikely to harm humans and pests cannot become resistant.

There are also some disadvantages to biological control. The predator/parasite does not kill all the pests; they work by controlling pest numbers, and keeping them at manageable levels. Also, the predator itself may become a problem in the environment. For example, cane toads were introduced into Australia to kill beetles (the pest), but there are now such a large number of cane toads that they are regarded as a pest.

Chemical control of pests involves using **herbicides**, **fungicides** or **insecticides** (together these are called pesticides) to kill the pests.

Advantages of chemical control are that the chemicals can kill all the pests and they are easy to use. Disadvantages include the fact that the chemicals are expensive and non-specific so they may kill other organisms as well as pests. Many pesticides are **persistent** and remain in the environment or the bodies of organisms for a long time. Pests may become resistant to them as a result of mutation, so they are no longer effective.

There is no single ideal method for killing pests. Chemical control is effective, but is likely to harm other organisms. Biological control is specific, but will not get rid of all pests. Integrated pest management is a combination of chemical, biological and cultural control.

4.6 Learning points

Summary

- Weeds compete with crop plants, while other pests and diseases damage crop plants, all of which reduce productivity.
- Properties of annual weeds include rapid growth, short life cycle, high seed output and long-term seed viability.
- Properties of perennial weeds include storage organs to provide a food source and the ability to reproduce by vegetative reproduction.
- Most of the pests of crop plants are invertebrate animals such as insects, nematode worms and molluscs.
- Pests feed on crop organisms and/or compete with crop organisms for resources, so yield is reduced.
- Pests can directly cause disease in crop organisms.
- Plant diseases can be caused by fungi, bacteria or viruses, which are often carried by invertebrates.
- Ploughing, weeding and crop rotation are cultural methods of pest control.
- The chemical control of pests involves using herbicides, fungicides, insecticides, molluscicides and nematicides - together these are called pesticides.
- Pesticides include herbicides to kill weeds, fungicides to control fungal diseases, insecticides to kill insect pests, molluscicides to kill mollusc pests and nematicides to kill nematode pests.
- Pesticides can be selective or systemic.
- Selective chemicals have a greater effect on certain plant species (broad leaved weeds).
- Systemic herbicide spreads through vascular system of plant and prevents regrowth.
- Systemic insecticides, molluscicides and nematicides spread through the vascular system of plants and kill pests feeding on plants.
- Protective applications of fungicide based on disease forecasts are often more effective than treating a diseased crop.
- Problems with pesticides include toxicity to non-target species, persistence in the environment, bioaccumulation or biomagnification in food chains, and the production of resistant populations of pests.
- Bioaccumulation is a build-up of a chemical in an organism.
- Biomagnification is an increase in the concentration of a chemical moving between trophic levels.
- In biological control the control agent is a natural predator, parasite or pathogen of the pest.

Summary continued

- There are risks associated with biological control, such as the control organism may become an invasive species, parasitise, prey on or be a pathogen of other species.

- Integrated pest management is a combination of chemical, biological and cultural control.

4.7 End of topic test

End of Topic 4 test Go online

Farmers growing soya beans have a problem because weeds compete with their crop. A genetically engineered variety of soya bean may solve their problem. A bacterial gene which can boost photosynthesis has been inserted into the plant. The new soya bean plants can withstand glyphosate, a herbicide, which disrupts photosynthesis and kills the plants.

Q3: A field of genetically engineered soya beans is sprayed with glyphosate. Explain the effects this would have on the yield.

....

Q4: Sometimes crop plants can interbreed with weeds. Explain one problem which could be caused if the genetically engineered soya beans did this.

....

Q5: The following table summarises the results from an investigation into the impact of a new pesticide on four pests.

Crop	Pest	Region of host attacked	Average loss of crop (acres/year) Without insecticide	Average loss of crop (acres/year) With insecticide
Apple	Aphid	Leaf and flower	12000	600
Pea	Weevil	Leaf and pod	4500	450
Potato	Leather jacket	Root and tuber	1800	1700
Cabbage	Caterpillar	Leaf and stalk	3200	400

The number of acres of pea crop saved per year by the use of the chemical was:

a) 450
b) 4005
c) 4050
d) 4500

....

Q6: On which crop did the chemical have the **greatest** effect relative to the others?

a) Apple
b) Pea
c) Potato
d) Cabbage

....

TOPIC 4. CROP PROTECTION

Q7: On which pest was the insecticide **least** effective?
a) Aphid
b) Weevil
c) Leather jacket
d) Caterpillar

....

Q8: Give one property of an annual weed.

....

Q9: The following table shows the concentration of a non-biodegradable pesticide residue in the tissues of the organisms in a food chain and the water in their ecosystem.

	Concentration of pesticide (ppm)
Water	0.00005
Plankton	0.04
Herbivorous fish	0.23
Carnivorous fish	2.07
Fish-eating birds	6.00

The concentration of pesticide increased by a factor of nine between:

a) herbivorous fish and carnivorous fish.
b) carnivorous fish and fish-eating birds.
c) plankton and herbivorous fish.
d) water and plankton.

....

Q10: The concentration of pesticide in the fish-eating bird is greater than that in the water by a factor of:

a) 1.2×10^3
b) 1.2×10^4
c) 1.2×10^5
d) 1.2×10^6

....

Q11: Complete the sentences using words from the list.

- Pesticides can cause problems in the environment because they can _____ within the body of an organism.
- They can also _____ along food chains.
- This means each successive organism in the food chain has a _____ concentration of the chemical in its tissues than the previous organism.

Word list: degrade, accumulate, concentrate, magnify, higher, lower.

....

Q12: The table below shows the energy content and amount of insecticide in the organisms of a food chain.

Organisms in food chain	Energy content as percentage of original energy	Amount of insecticide in body mass (mg/kg)
Human	1	1.0
Fish	4	0.1
Microscopic animals	20	0.01
Microscopic plants	100	0.001

Calculate the percentage loss of energy between the microscopic water plants and microscopic animals.

....

Q13: How are insecticides useful to farmers?

....

Q14: Explain why fish contain insecticides.

....

Q15: The use of pesticides may result in a population selection pressure producing a _____ population.

....

Q16: Which control methods are used in integrated pest management?

a) Chemical and cultural
b) Biological and chemical
c) Cultural and biological
d) Chemical, cultural and biological

Unit 3 Topic 5

Animal welfare

Contents

5.1 Animal welfare 314
5.2 Behavioural indicators of poor welfare 316
5.3 Learning points 318
5.4 End of topic test 318

Learning objective

By the end of this topic you should be able to:

- describe the costs, benefits and ethics of providing different levels of animal welfare in livestock production;
- describe behavioural indicators of poor welfare;
- describe stereotype behaviour, misdirected behaviour, failure in sexual and parental behaviour and altered levels of activity.

5.1 Animal welfare

Animal welfare refers to both the physical and mental well-being of animals. The term animal welfare can also mean human concern for animal health. Animal welfare is measured by indicators including behaviour, physiology, longevity, and reproduction.

In the 1960s, the British Government commissioned the Brambell Report on intensive animal production. Intensive animal production refers to large farms raising a great number of animals in small spaces compared to traditional farms. The Brambell committee listed the five freedoms animals should have. The first three refer to physical welfare and the last two refer to mental welfare.

1. *Freedom from hunger and thirst* - the animal should be able to drink fresh water whenever they need it and they should be fed on a diet which keeps them healthy and strong.
2. *Freedom from chronic discomfort* - the animals should be kept in a comfortable environment. The animals should not be too hot in summer or too cold in winter, there should be plenty of fresh air and they should have a comfortable, dry place to lie down.
3. *Freedom from pain, injury and disease* - the environment that the animal lives in shall be safe for them and not cause them injury. If the animals have any problem of injury or disease, a vet should be called immediately.
4. *Freedom to express normal behaviour* - the animals should be able to move around easily and mix with others in their group.
5. *Freedom from fear and the avoidance of stress whenever possible* - the animals should not be kept in conditions where they are afraid or where they might suffer any unnecessary pain or distress. This also applies when they are in transport, at market or abattoirs.

Most farming practices have been developed over the years to meet the demand from the public for sufficient supplies of food at relatively low cost. However, some of these practices may be considered by some people to be inappropriate for the welfare of the animals involved.

Ensuring farm animals are given the five freedoms detailed above can be costly for farmers; however they can give long-term benefits. Animals which have experienced good welfare have a better growth rate, increased reproductive success and produce products of a higher quality than animals which have experienced poor welfare. The UK maintains some of the highest animal welfare standards in Europe; while this may provide benefits, such as increased growth rate, it means animal products produced in the UK are more expensive than those from other parts of Europe.

When considering **animal welfare** issues, ethical considerations must also be taken into account. Battery reared chickens present an example of an ethical issue surrounding animal welfare. The chickens are kept in small cages with sloping mesh floors - the eggs roll down into a channel to be collected and the droppings fall through the mesh. The chickens have limited movement and often show unusual behaviours such as feather pecking. Rearing chickens in this way is much cheaper than rearing free-range chickens, however, can we justify this treatment of chickens to ensure we have a plentiful and cheap supply of eggs?

In general, intensive farming is less ethical than free range farming due to poorer animal welfare. Intensive farming often creates conditions of poor animal welfare but is usually more cost effective, generating higher profit as costs are low. Free range requires more land and is more labour intensive, but products produced in this way can be sold at a higher price and animals have a better quality of life.

TOPIC 5. ANIMAL WELFARE

Battery reared chickens (http://en.wikipedia.org/wiki/File:Battery_hens_-Bastos,_Sao_Paulo,_Braz il-31March2007.jpg by http://commons.wikimedia.org/wiki/User:Snowmanradio, licensed under http ://creativecommons.org/licenses/by/2.0/deed.en)

Animal welfare: Animal freedoms Go online

Q1: Complete the following table with the correct examples of freedom.

Freedoms for Animals	Example
Freedom from hunger and thirst.	
Freedom from chronic discomfort.	
Freedom from pain, injury and disease.	
Freedom to express normal behaviour.	
Freedom from fear and the avoidance of stress whenever possible.	

Examples:

- Animals should be able to move around freely and mix with other animals in the group.
- Animals should not be exposed to unnecessary pain.
- Environment should be safe for animals and not cause them injury.
- Animals should be able to drink fresh water when they need it.
- Animals should be kept in a comfortable environment.

5.2 Behavioural indicators of poor welfare

When animals experience poor welfare standards they may display certain unusual/uncharacteristic behaviours. The behavioural indicators of poor welfare include the following:

- **stereotype**;
- **misdirected behaviour**;
- failure in sexual and parental behaviour;
- altered levels of activity.

A stereotype is a behaviour which involves unusual repetitive movement quite unlike those shown by wild animals. These behaviours are invariant (performed over and over again in the same way) and have no obvious function. The most common types of stereotypies can be grouped in the following ways:

- pacing-type - constantly walking back and forth or walking the same circuit around an enclosure;
- oral-type - among these are obsessive object licking and tongue rolling. Oral stereotypes are common in all grazing animals. Sows confined in stalls may repeatedly rub their mouths backwards and forwards over the bars, even making themselves bleed;
- others - include rocking backwards and forwards or side to side, rubbing continually against an object, head shaking, eye rolling (seen in veal calves), sham chewing for hours day after day (seen mainly in tethered sows) and tongue rolling (seen often in cows).

Misdirected behaviour

The freedom to express normal behaviour for domesticated animals is complicated and hard to apply in the real world. In many cases it is impossible to give a domesticated or captive animal the freedom to express a normal behaviour. The most effective indicator of suffering is the observation of abnormal misdirected behaviour in confined animals.

Misdirected behaviour involves a normal behaviour being displayed in a different/inappropriate situation. This behaviour may be misdirected towards the individual itself, other members of its species or its surroundings. Abnormal misdirected behaviour comes in a variety of forms. Examples are:

- feather pecking amongs battery hens;
- tail biting in pigs;
- chewing cage bars or other inanimate objects in pigs.

All of these misdirected behaviours are clear signs of suffering by the animal. They can be reduced by providing animals with members of its own species to interact with in a large stimulating enclosure.

Battery chicken recovering outside after feather pecking

Failure in sexual or parental behaviour

Reproductive success can be used as a measure of **animal welfare**. Those animals kept in low welfare conditions often have low reproductive success; these animals may fail to perform sexual behaviour or may reject/neglect any offspring they do produce. For an animal to develop normal sexual and parental behaviour it must be allowed to interact with members of its own species in a suitable enclosure.

Altered levels of activity

Altered levels of activity are also behavioural indicators of poor welfare. Examples are:

- apathy, particularly apparent in sows confined in stalls;
- hysteria that occurs among chickens and turkeys which can lead to the birds piling on top of each other, thereby causing death.

5.3 Learning points

Summary

- The costs of providing different levels of animal welfare in livestock production are financial.
- The benefits of good animal welfare include faster growth rate, better quality products and increased reproductive success.
- There are also ethical considerations which must be taken into account when providing different levels of animal welfare in livestock production.
- Intensive farming is less ethical than free range farming due to poorer animal welfare.
- Intensive farming often creates conditions of poor animal welfare but is often more cost effective, generating higher profit as costs are low.
- Free range requires more land and is more labour intensive, but can be sold at a higher price and animals have a better quality of life.
- There are behavioural indicators of poor welfare. These behavioural indicators include:
    - stereotypes of behaviour patterns - a repetitive movement;
    - misdirected behaviour - a normal behaviour displayed in a different/inappropriate situation;
    - failure in sexual and parental behaviour;
    - altered levels of activity.

5.4 End of topic test

End of topic 5 test Go online

Q2: The following list describes observed behaviour of pigs on a farm.

1. Repeated wounding of other pigs by biting.
2. Sham chewing for hours.
3. Biting of own tail.

Which of these behaviours indicate poor animal welfare?

a) 1 and 2 only
b) 1 and 3 only
c) 2 and 3 only
d) 1, 2 and 3

....

Q3: The following are all examples of behaviour patterns exhibited by animals.
1. Sows rubbing mouth backwards and forwards over bars of cage
2. Lions pacing backwards and forwards on the exact same path
3. Chickens pecking their own feathers out
4. Red jungle fowl scratching and pecking food from soil surfaces

Which of the above are examples of stereotypic behaviour?

a) 1 and 2
b) 1 and 3
c) 2 and 4
d) 3 and 4

....

Q4: Complete the following table by matching the examples of abnormal behaviour with the types.

Type of abnormal behaviour	Example of abnormal behaviour
Stereotype behaviour	
Misdirected behaviour	
Failure in sexual behaviour	
Altered levels of activity	

Examples:
- Hysteria among turkeys
- Tail biting in pigs
- Polar bears pacing in a zoo
- Cheetahs unable to breed in captivity

Unit 3 Topic 6

Symbiosis

Contents

6.1 Symbiosis 323
6.2 Parasitism 324
6.3 Malaria 325
6.4 Mutualism 326
6.5 Learning points 328
6.6 Extended response question 328
6.7 End of topic test 329

Learning objective

By the end of this topic you should be able to:

- state that symbiotic relationships are coevolved, intimate relationships between members of two different species;
- know that a parasite benefits in terms of energy or nutrients;
- know that the parasite's host is harmed by the loss of these resources;
- understand that parasites often have limited metabolism so often cannot survive without a host;
- know that transmission of parasites to new hosts involves one of the following:
  - using direct contact;
  - resistant stages;
  - vectors;
- understand that evolution of parasitic lifestyles involves intermediate (secondary) hosts;
- state that in mutualism both partner species benefit in an interdependent relationship.

6.1 Symbiosis

Symbiosis refers to a relationship between two organisms from different species. Symbiotic relationships involve direct contact between members of the two species and have usually evolved over millions of years. In symbiosis, both individuals usually show adaptations which allow the relationship to take place.

In many symbiotic associations, one partner lives inside the other. For example, a variety of micro-organisms, mainly bacteria, live in the alimentary canal of animals such as cows. The bacteria provide the enzyme cellulase to aid the breakdown of cellulose into simple sugars which the cow uses for respiration. In return the bacteria are provided with enough food to live and a suitable environment in which to grow.

A well-known example of a symbiotic relationship is found in lichens. Lichen is a composite organism formed as a result of a union between a green alga and a fungus. The fungus gains oxygen and carbohydrates from the alga, whilst the alga obtains water, carbon dioxide and mineral salts from the fungus as well as protection from drying out. The result of this symbiotic relationship is an extremely hardy organism which can survive in extreme places, for example on exposed rocks at high altitudes and in Arctic and Antarctic regions.

Lichen

Lichens provide an example of an extremely well balanced relationship. The two partners do nothing but good for one another. However, not all symbiotic partnerships are as harmonious as this. In some cases the benefits enjoyed by one of the participants may be marginal, and in others the relationship may be detrimental to one participant.

A tapeworm is a type of flatworm which is capable of living in the guts of animals such as humans, cows and pigs. Once inside the gut of an animal (the parasite's host), the tapeworm is provided with a warm environment and a plentiful supply of food. This is detrimental to the host which loses nutrition and suffers weight loss.

Symbiotic relationships can be grouped into one of two categories:

- Parasitic - where one organism benefits from the relationship and the other is harmed;
- Mutualistic - where both organisms benefit from the relationship.

6.2 Parasitism

Parasitism refers to a relationship where one species benefits at the expense of the other. A parasite benefits in terms of energy or nutrients, whereas its host is harmed by the loss of these resources. Parasites often have a limited metabolism so cannot survive out of contact with its host.

In a parasitic relationship, the host species is always exploited to some degree, although often in such a way that its health is impaired only slowly. This allows the parasite to exploit its host over a longer period of time.

The transmission of parasites to new hosts can occur by the following methods:

- Using direct contact;
- Resistant stages;
- Vectors.

Using direct contact

Some parasites spread from one host to the next by physical contact. For example head lice pass from one person to another by direct contact. Living with our furry friends exposes us to many canine and feline parasites, for example dogs can transmit 65 types of parasites to humans through direct contact.

Human head louse - spread by direct contact

Resistant stages

Some parasites have resistant stages in their lifecycle. Resistant stages are a part of the parasites lifecycle where they are resistant to adverse environmental conditions and can survive for long periods of time.

The human tapeworm has resistant stages in its life cycle. *Taenia solium* is the human tapeworm which can be contracted from pork. The pig becomes infected from ingesting the eggs and once inside the intestine, the eggs release the oncosphere (first-stage larvae). The oncosphere then migrates to the muscles, where it develops into a cyst-like structure. The cyst can survive for several years in the tissue of the pig; this is the resistant stage of the lifecycle. Humans become infected by ingesting raw or undercooked infected meat. When a human becomes infected by tapeworms they can suffer from symptoms such as stomach pains, vomiting and weight loss.

TOPIC 6. SYMBIOSIS

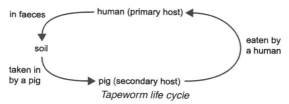

Tapeworm life cycle

Many parasites use only one type of host organism for all stages of their life cycle for example the cat flea. Some parasites have developed the use of intermediate (secondary) hosts, as describe in the example shown above. In this case part of the lifecycle is spent in a different organism, known as an intermediate (secondary) host. In most cases this is beneficial to the parasite as it increases the chance of its offspring being transmitted to the primary host and many parasites are capable of increasing in numbers inside their intermediate (secondary) host by asexual reproduction.

Vectors

A vector is a carrier which allows a parasite to pass from one host to another. The most common example of a vector is the mosquito which carries the *Plasmodium* protozoa (which causes malaria) from human to human.

6.3 Malaria

Malaria is caused by a unicellular parasite known as *Plasmodium*. This enters the body after a bite from an infected mosquito. The mosquito acts as the vector which carries the malarial parasite. The *Plasmodia* invade red blood cells and multiply inside them. When they burst out of the red blood cells, toxins are released and the person develops a fever. The *Plasmodia* then infect more red blood cells and liver cells, causing serious damage. If the infected person is bitten by another mosquito which sucks their blood, *Plasmodia* are transferred to another mosquito ready to infect another human. In this example, mosquitoes act as vectors; they do not suffer from the disease, but they carry it from one person to another.

Malaria is a serious disease which kills about 2 million people each year. Incidences of malaria can be greatly reduced by:

- preventing mosquito bites using insect repellents or mosquito nets;
- killing the *Plasmodium* in the body by using anti-malarial drugs such as quinine;
- killing mosquitoes with insecticides;
- preventing mosquitoes breeding by draining swamps.

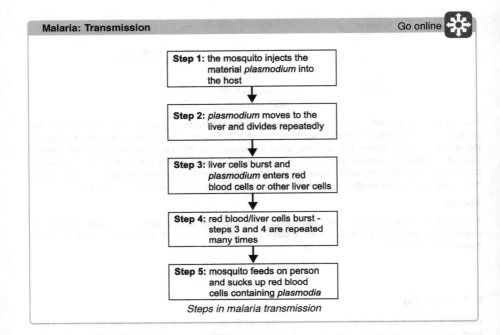
Steps in malaria transmission

6.4 Mutualism

Mutualism describes a relationship where both participants benefit from the interaction. Mutualistic relationships are described as interdependent, this means one cannot live without the other. Two examples of mutualistic relationships are outlined below.

Cellulose-digesting protozoa/bacteria in the guts of many herbivores

Animals such as cows feed by grazing on plant matter. This plant matter contains large quantities of cellulose which makes up the cell wall of plant cells. In order to break down cellulose, the enzyme cellulase is required. Cows are not capable of producing this enzyme; therefore they rely on a mutualistic relationship with the micro-organisms (mainly bacteria and protozoa) in their gut to help digest their food.

The micro-organisms found in parts of a cow's stomach produce the enzyme cellulase which allows cellulose to be broken down into its constituent sugars. These sugars can be used by the cow as a source of energy. Other metabolites are also produced which are useful for the cow. In return the micro-organisms are provided with a warm and safe place to live, which has a plentiful food supply.

Photosynthetic algae in the polyps of corals

Coral reefs rely heavily on mutualistic **symbiosis**. The living coral organisms are animals called polyps, close relatives of the jellyfish. Each polyp lives in a small cup-like skeleton of calcium carbonate that it secretes itself, and over thousands of years, the layers of calcium carbonate build up to form coral reefs. Like jellyfish they use feathery tentacles to capture crustaceans and other small animals.

Coral polyps

Polyps also depend on the energy provided by single-celled algae (*Zooxanthellae*) which they shelter within and between their cells. The algae carry out photosynthesis to produce carbohydrates. These carbohydrates are used by both the algae and the polyp as a source of energy. In return, the algae are provided with a sheltered place to live and a supply of nitrogen compounds from the polyps wastes which can be used to produce proteins.

6.5 Learning points

Summary

- Symbiotic relationships are described as co-evolved, intimate interactions between the members of two different species.
- In parasitism, one organism benefits whilst the other is harmed by the interaction.
- A parasite benefits in terms of energy or nutrients, whereas its host is harmed by the loss of these resources.
- Parasites often have limited metabolism so often cannot survive without a host.
- Normally a balance exists between parasitic damage and host defense, resulting in a relatively stable relationship.
- Parasites may be transmitted between hosts by vectors, direct contact or through resistant stages.
- In mutualism, both organisms benefit from the interaction.
- Organisms in mutualistic relationships exchange metabolites and are structurally compatible.

6.6 Extended response question

The activity which follows presents an extended response question similar to the style that you will encounter in the examination.

You should have a good understanding of mutualism and parasitism before attempting the question.

You should give your completed answer to your teacher or tutor for marking, or try to mark it yourself using the suggested marking scheme.

Extended response question: Mutualism and parasitism

Discuss interactions between species under the following headings:

A) Mutualism *(2 marks)*

B) Parasitism *(4 marks)*

6.7 End of topic test

End of Topic 6 test Go online

Q1: Malaria is a disease which kills many people in tropical parts of the world. The female mosquito acts as a vector for the disease. What is meant by the term vector? (*2 marks*)

....

Q2: The following diagram shows the life cycle of the mosquito.

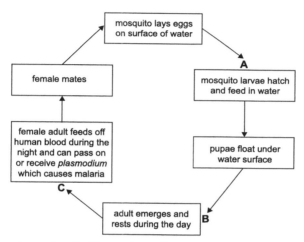

Match the correct control method to each lettered stage on the diagram.

- Add fish to the water to eat eggs or larvae/drain wet areas/spray water with insecticide.
- Use insecticides to kill adults/use mosquito nets or repellent to prevent being bitten.
- Add oil or detergent to the water surface to stop pupae breathing and prevent adults emerging.

....

Q3: A species of ant which is found in Latin America inhabits the thorns of a tropical shrub know as Acacia. The ants receive nectar and shelter from the plant. The plant receives protection from the ants.

This is an example of:

a) parasitism.
b) predation.
c) grazing.
d) mutualism.

....

Q4: Hydra is a small freshwater animal that uses its tentacles to catch food. One variety (green hydra) has photosynthetic algae living in its tissues. Another variety (colourless hydra) has no algae.

The relationship between Hydra and the algae is believed to be an example of mutualism.

Under what conditions would a comparison of the growth rates of green and colourless Hydra test this hypothesis?

a) Light; food supplied
b) Light; no food supplied
c) Dark; food supplied
d) Dark; no food supplied

....

Q5: Parasites may be transmitted between closely related species. State one way in which parasites can be transmitted.

....

Q6: Symbiosis is the term used to describe close interactions between organisms. For each of the following statements, indicate which type of symbiosis it describes.

1. One species benefits and the other is harmed.
2. Both species in the interaction benefit.

....

Q7: The following table shows four examples of interactions between species. Which column in the table shows correctly the benefits (+) or costs (-) which result from each interaction?

Interaction	A	B	C	D
Sheep grazing in a field of grass	+/-	+/-	+/+	+/-
Owls and foxes hunting for the same food	+/-	-/-	-/-	+/-
Corals acting as hosts for *zooxanthellae*	+/-	+/+	+/-	+/+
"Cleaner fish" feeding on parasites which they remove from other fish	+/+	+/+	+/-	+/+

....

Q8: Adult beef tapeworms live in the intestine of humans. Segments of the adult worm are released in the faeces, and embryos which develop from them remain viable for five months. The embryos may be eaten by cattle and develop in their muscle tissue. Which row in the following table correctly identifies the various roles in the tapeworm life cycle?

	Role of human	Role of embryo	Role of cattle
A	Intermediate (secondary) host	Resistant stage	Vector
B	Intermediate (secondary) host	Vector	Host
C	Primary Host	Vector	Intermediate (secondary) host
D	Primary Host	Resistant stage	Intermediate (secondary) host

Q9: The tentacles of coral polyps have a symbiotic relationship with photosynthetic algal cells. The coral polyps provide shelter to the algal cells and the algal cells provide the coral polyps with a source of energy in the carbohydrates they produce.

What type of symbiosis is found in the relationship between coral polyps and photosynthetic algae?

Unit 3 Topic 7

Social behaviour

Contents

7.1 Social behaviour 335
7.2 Social hierarchy 335
7.3 Cooperative hunting 338
7.4 Social mechanisms for defence 339
7.5 Altruism and kin selection 340
7.6 Social insects 344
7.7 Primate behaviour 345
7.8 Learning points 346
7.9 Extended response question 348
7.10 End of topic test 349

> **Learning objective**
>
> By the end of this topic you should be able to:
>
> - understand that many animals live in groups and have behaviour that is adapted to group living;
> - state that adaptations to living in groups include social hierarchy, cooperative hunting and defence;
> - describe the features of social hierarchy;
> - describe the features and benefits of cooperative hunting;
> - give examples of social defence in animals;
> - describe the features of altruistic behaviour;
> - identify examples of altruistic behaviour;
> - explain reciprocal altruism;
> - understand that behaviour that appears to be altruistic can be common between a donor and a recipient if they are related (kin);
> - describe the features and advantages of kin selection;
> - describe the structure of insect societies such as bees, wasps, ants and termites;
> - give examples of primate behaviours that support social structure to reduce unnecessary conflict;
> - give examples of primate group behaviour;
> - state that the long period of parental care in primates gives an opportunity to learn complex social behaviours.

TOPIC 7. SOCIAL BEHAVIOUR

7.1 Social behaviour

Many animals live in social groups and have behaviours that are adapted to group living. Group members use social signals to establish behaviours which benefit both individuals and the group as a whole. The table below outlines some benefits and costs of living in a social group.

Benefits	Costs
Individual risk of predation diluted by joining a group	Greater risk of contracting disease
Groups can tackle larger prey than individuals	Greater chance of mistakenly feeding someone else's offspring
Grouping confuses predators, making it harder for them to target prey	Investment in foraging, courtship, or other activities exploited by other group members
Huddling in groups help thermoregulation	Young may be cannibalised by neighbours
Energetic advantages to swimming or flying in a group through 'slipstreaming'	Greater risk of inbreeding

Social grouping

7.2 Social hierarchy

Within a group of animals, a social hierarchy is often found to operate. A social hierarchy is a rank order within a group of animals consisting of a dominant and subordinate members. In a social hierarchy, dominant individuals carry out ritualistic (threat) displays whilst subordinate animals carry out appeasement behaviour to reduce conflict. Animals often form alliances in social hierarchies to increase their social status within the group. Social hierarchies benefit species as they increase the chances of the dominant animal's favourable genes being passed on to offspring.

Examples of social hierarchy are often seen in groups of newly hatched birds where one will soon emerge as the dominant member of the group. This bird is able to peck and intimidate all other members of the group without being attacked in return. It therefore gets first choice of any available food. Below this dominant bird there is a second one which can peck all others except the first and so on down the line. This linear form of social organisation is called a **pecking order**.

The following table summarises the results from observing a group of newly hatched chickens over a period of time. Chicken A dominates all of the others, B dominates all of the others except A and so on down the line to chicken H.

		Chicken receiving pecks						
		A	B	C	D	E	F	G
Chicken giving pecks	A		√	√	√	√	√	√
	B			√	√	√	√	√
	C				√	√	√	√
	D					√	√	√
	E						√	√
	F							√
	G							
	H							

Pecking order in chickens

Social hierarchies can also be observed in groups of mammals such as wolves and baboons. Social signals, such as aggressive behaviour, are used to establish a rank order within the group with the most dominant individual getting first choice of food, preferred sleeping places and available mates.

Dominant behaviour (http://en.wikipedia.org/wiki/File:Wolves_Kill.jpg by http://commons.wikimedia.org/wiki/User:Ipuser, licensed under http://creativecommons.org/licenses/by/2.0/deed.en)

Social hierarchy: Questions

Five male zebra finches, P, Q, R, S and T, were kept together and observed over a period of several days. During this time, a record was kept of the results from 20 confrontations between each pair of birds. The bird which successfully dominated its rival in each contest was given a score of one point.

Q1: The results are shown in the following table.

Contest	Score out of 20 (points)	Winner	Net number of contests won
T v Q	T 17, Q 3	T	14
T v R	T 3, R 17		
P v Q	P 18, Q 2		
Q v R	Q 0, R 20		
Q v S	Q 8, S 12		
R v P	R 13, P 7		
P v T	P 14, T 6		
S v T	S 5, T 15		
R v S	R 19, S 1		
S v P	S 4, P 16		

Complete the two right-hand columns in the table. (The first example has been done for you.)

....

Q2: Which bird has the lowest status and is at the bottom of the pecking order? Explain your choice.

....

Q3: Which bird is at the top of the dominance hierarchy? Explain your answer.

....

Q4: Give the pecking order of the five birds.

7.3 Cooperative hunting

Cooperative hunting involves a group of animals working together to find and catch prey. Cooperative hunting benefits all members of a social group, including lower ranking individuals as the subordinate animals may gain more food than by foraging alone. Food sharing will occur as long as the reward for sharing exceeds that for foraging individually.

Many animals rely on cooperation between individuals to catch prey. Dolphins often work in groups to catch their prey, one dolphin (the driver) herds a shoal of fish towards the rest of the dolphin group who form a barrier. The fish are forced to leap out of the water into the air, where the dolphins catch them.

Dolphins: cooperative hunting

Lions hunt in groups of three to seven and perform one of two roles, centre or wing. The lions on the edge of the group are the wings, they run around their target and drive it towards the centres who are lying waiting to ambush the prey. The centres leap up and attack the prey; the kill is then shared amongst the group.

Advantages of cooperative hunting

The main advantage of cooperative hunting lies in the fact that the kill is shared between all members of the group. This means that even those organisms at the bottom of the social hierarchy (subordinate individuals) obtain food. Group hunting also allows larger prey animals to be hunted meaning all individuals gain more food than they would by foraging alone. One final benefit of cooperative hunting is that by working as a group, less energy is used per individual in obtaining the prey, this maximizes energy gain.

In the case of lions, an additional advantage to group hunting lies in the fact that after the kill, there will be more individuals to keep scavengers and other potential thieves away from the carcass. Therefore, the defence of the kill is also a benefit of cooperative hunting.

7.4 Social mechanisms for defence

Many animals live in social groups, not only to benefit from increased access to food, but also for defence. Social defence is often thought of as "safety in numbers" and can operate in several different ways. In some species having many individuals means the group is able to fend off attacks by predators. In other species it means there are more eyes looking out for danger and the group can take cover when alerted to the presence of a predator.

Colonial nesting birds - gulls and terns, for example - may provide formidable opposition to an invading predator such as a fox by mobbing it, even hitting the predator with their feet.

Terns

Even though each bird is responding individually to defend its own nest, the proximity of other birds all doing the same thing means that their combined efforts can be much more effective than that of a single bird on its own. As a result, the nesting success of gulls in a large colony is considerably greater than that of gulls that nest singly or in small groups.

In meerkats, social defence can be seen by particular individuals who undertake vigilance duties and take turns to go to a high look-out point such as a tree and keep watch for predators while the others feed. If the area is safe the sentry makes quiet peeping sounds. When the sentry sees a predator, he barks loudly and all the meerkats in the group retreat into their burrows.

Meerkat lookout

Some species of antelope live on open grassy plains with very little cover. As well as being able to run away from predators, some species encircle their young when under attack. This means predators must risk being gored by the antelopes' horns to get their prey and vulnerable individuals (who are likely to be caught) are protected.

Social mechanisms for defence: Questions Go online

An experiment was set up to investigate the attacks of a predatory bird, a goshawk, on pigeons. The attack success of the goshawk against different sized flocks of pigeons was studied. The results are shown in the table.

Number of pigeons in flock	Percentage attack success of goshawk
1 - 5	80
6 - 10	60
11 - 49	18
50 +	10

Q5: Using the information from the table, draw a bar-graph for the results obtained.

....

Q6: Calculate the percentage decrease in percentage attack success from a flock of pigeons of numbering 1-5 and a flock of pigeons numbering 50+.

....

Q7: Describe what the results of the experiment show.

....

Q8: Suggest a possible explanation for the results.

7.5 Altruism and kin selection

Animals usually show behaviours which are beneficial to their own chances of survival. In some cases an animal will behave in a manner which is harmful to itself but beneficial to another individual. This behaviour is described as altruism. An example of altruism can be seen in wolves which bring meat back to members of the group who were not present at the kill. The "donor" wolf must expend energy carrying the kill back and the "recipient" wolf benefits as it gains access to food.

The social insects, such as termites, ants, bees and wasps, show extreme altruistic behaviour. There is usually just one reproductive female (the queen) and large numbers of sterile workers. The workers perform all the tasks of the society such as foraging, rearing young, nest construction and defence, and do not reproduce at all themselves.

A queen bee (centre) with attendants (http://commons.wikimedia.org/wiki/File:Adult_queen_bee.jp g by http://en.wikipedia.org/wiki/User:Pollinator, licensed under http://creativecommons.org/license s/by-sa/3.0/deed.en)

Donor vs recipient

Reciprocal altruism is a behaviour whereby an organism (donor) acts in a manner that temporarily reduces its fitness while increasing another organism's (recipient) fitness with the expectation that, the roles of donor and recipient later reverse. Grooming in primates can be thought of as an example of reciprocal altruism. An individual will expend time and energy grooming another member of the group in the expectation that the favour will be returned in the future.

A typical group of vampire bats exhibit reciprocal altruism by regurgitating blood meals to other bats. Although members of the group are largely unrelated, they share their meal with others in the group. To avoid starvation, vampire bats require frequent blood meals. Individuals often regurgitate part of their blood meal to other bats, but they are more likely to do so for those who have shared a meal with them in the past. Therefore, it can be seen that vampire bats employ some sort of reciprocal altruistic strategy when it comes to blood sharing.

Behaviour that appears to be altruistic can be common between a donor and a recipient if they are related (kin). The donor will benefit in terms of the increased chances of survival of shared genes in the recipient's offspring or future offspring.

Favourable conditions for kin-selected cooperation are widespread. Around 96% of birds and 90% of mammals that live in family groups show cooperative breeding by helping others to rear offspring. There is also evidence that helping is directed towards more closely related recipients within family groups. For example, in white-fronted bee-eaters, helpers often chose to aid the pair they themselves were most closely related to.

White-fronted bee-eater (https://en.wikipedia.org/wiki/File:Merops_bullockoides_1_Luc_Viatour.jpg by https://commons.wikimedia.org/wiki/User:Lviatour, licensed under https://creativecommons.org/licenses/by-sa/3.0/deed.en)

Altruism and kin selection: Questions Go online

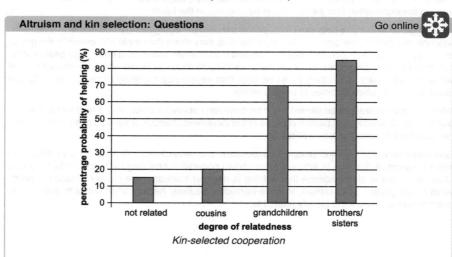

Kin-selected cooperation

Q9: From the graph, calculate the percentage increase in the degree of relatedness in white-fronted bee-eaters between cousins and grandchildren.

....

Q10: Calculate the percentage for the overall percentage probability of helping relatives as opposed to non-relatives.

....

Q11: What conclusion can you come to from the results of the graph?

Q12: Which of the terms listed match with the following definitions?

- A behaviour in which an organism acts in a manner that temporarily reduces its fitness while increasing another organism's fitness with the expectation that the other organism will act in a similar manner later on.
- Society in which some individuals are dominant to others who are submissive to the dominant ones.
- Strategies in evolution that favour the reproductive success of an organism's relatives even at the cost of an organism's own survival and reproduction.

Terms: Kin selection, Reciprocal altruism, Social hierarchy.

7.6 Social insects

The evolution of the societies of insects can be seen in such social insects as bees, wasps, ants and termites, in which only some individuals contribute to reproduction. The rest of the group is involved in gathering food and defending the colony. This benefits the species as a whole, because the "workers" become specialised in performing their function; although they are not directly involved in the reproductive process, the tasks they complete ensure the survival of the species.

The "workers" can be thought of as a rank within the insect society. One of the most important factors determining rank is what the insects are fed when young. In bees, wasps and termites, all eggs laid by the queen are potentially equal, but most larvae are fed a restricted diet and develop into workers. Only richly fed individuals develop into the reproductive rank.

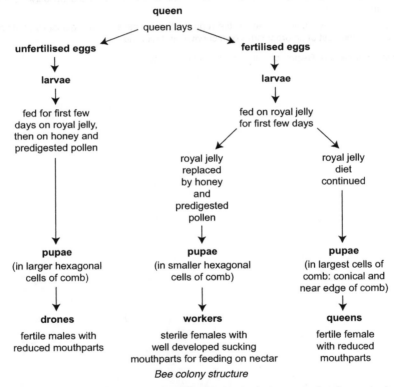

Bee colony structure

In the honey bee colony the queen is solely responsible for laying eggs, the drones for fertilising her, and the workers for gathering food and performing sundry duties in the hive such as defending the hive, collecting pollen and carrying out waggle dances to show the direction of food. Each rank is adapted for its particular job: thus the queen is the fertile female, the drones fertile males, and the workers sterile females with well-developed mouth parts and other structural adaptations for collecting nectar and pollen.

7.7 Primate behaviour

In primates, dominance and subordination are important features of their relationship with others. Dominance always involves the threat of physical displacement or attack, even though it is rarely observed once rank is established. It is important to remember that relationships within primate groups are as much characterised by positive interactions as by negative ones.

There are many friendly contacts between primates, such as moving and resting together, inviting grooming or offering to groom another. Mutual grooming is very important as a **placatory** gesture in primates. Often a dominant animal will allow itself to be groomed by a subordinate following a brief threat to which the subordinate has deferred.

To reduce unnecessary conflict, social primates use ritualistic display and appeasement behaviours. Ritualistic display involves one individual asserting its dominance over another by displaying aggressive behaviour. Appeasement behaviours include grooming, submissive facial expressions/body posture and sexual presentation. Sexual presentation as an appeasement gesture is very common in baboons and chimpanzees and is often made by females towards a dominant male.

The social **hierarchies** found within primate groups are complex and subject to change. In some monkeys and apes, alliances form between individuals which are often used to increase social status within the group.

Vervet monkeys grooming (https://www.flickr.com/photos/wwarby/2404512619/ by https://www.flickr.com/photos/wwarby/, licensed under http://creativecommons.org/licenses/by/2.0/deed.en)

Primate groups such as lemurs have been extensively researched. There is a social hierarchy within the troop which allows it to work well as a cohesive unit. Lemurs have group territories within their woodland habitat whose boundaries are very stable. They are marked by scent, and are defended by calling, which is usually sufficient to cause a neighbouring troop to retreat without further threat or fighting.

There is always close contact between a mother lemur and her infant, who clings continuously to her at first and is carried around everywhere. As it grows older other adults approach and play with the infant. This long period of parental care in primates gives an opportunity to learn complex social behaviours and to establish networks among other individuals in the troop. Lemurs have thick, dense fur and groom frequently. Mothers groom their infants and adults frequently groom each other.

Lemur

7.8 Learning points

Summary

- Many animals live in social groups and have behaviour that is adapted to group living.
- Social **hierarchy** is a rank order within a group of animals consisting of a dominant and subordinate members.
- In a social hierarchy, dominant individuals carry out ritualistic (threat) displays whilst subordinate animals carry out appeasement behaviour to reduce conflict.
- Social hierarchies increase the chances of the dominant animal's favourable genes being passed on to offspring.
- Animals often form alliances in social hierarchies to increase their social status within the group.
- **Cooperative hunting** may benefit subordinate animals as well as dominant.
- By cooperative hunting, large prey can be killed which would prove impossible for solitary animals.
- Subordinate animals may gain more food than by foraging alone.
- Food sharing will occur as long as the reward for sharing exceeds that for foraging individually.
- Co-operative hunting enables larger prey to be caught and increases the chance of success.
- Social defence strategies increase the chance of survival as some individuals can watch for predators whilst others can forage for food. Groups adopt specialised formations when under attack protecting their young.
- Altruistic behaviour harms the donor individual but benefits the recipient.

TOPIC 7. SOCIAL BEHAVIOUR

Summary continued

- Reciprocal altruism is a behaviour whereby an organism acts in a manner that temporarily reduces its fitness while increasing another organism's fitness with the expectation that the other organism will act in a similar way later on.
- Reciprocal altruism often occurs in social animals.
- Behaviour that appears to be altruistic can be common between a donor and a recipient if they are related (kin).
- Kin selection involves strategies that favour the reproductive success of an organism's relatives even at a cost to an organisms own survival and reproduction.
- The donor will benefit in terms of the increased chances of survival of shared genes which can be passed to recipient's offspring or future offspring.
- Social insects such as bees, wasps, ants and termites have a social structure where only some individuals contribute reproductively; most members of the colony are workers who cooperate with close relatives to raise relatives.
- Other examples of workers' roles include defending the hive, collecting pollen and carrying out waggle dances to show the direction of food.
- Sterile workers raise relatives to increase survival of shared genes.
- Primates display complex behaviours that support social structure to reduce unnecessary conflict.
- To reduce unnecessary conflict, social primates use ritualistic display and appeasement behaviours including grooming, facial expression, body posture and sexual presentation.
- Long period of parental care in primates gives an opportunity to learn complex social behaviours.
- In some monkeys and apes, alliances form between individuals which are often used to increase social status within the group.

7.9 Extended response question

The activity which follows presents an extended response question similar to the style that you will encounter in the examination.

You should have a good understanding of social behaviour before attempting the question.

You should give your completed answer to your teacher or tutor for marking, or try to mark it yourself using the suggested marking scheme.

Extended response question: Social behaviour

Write notes on social behaviour under the following headings:

A) Altruism and kin selection (*5 marks*)

B) Primate behaviour (*5 marks*)

7.10 End of topic test

End of Topic 7 test — Go online

Q13: A pack of African wild dogs catches large prey animals (such as wildebeest) by running it down to the point of exhaustion. Give two advantages gained by the dogs from this form of cooperative hunting.

....

Q14: Hawks are predators which attack flocks of pigeons. The table shows how the percentage of attack success of a predatory hawk varies with the number of pigeons in the flock.

Number of pigeons in the flock	% attack success
2	80
10	50
20	40
40	15

1. Calculate the percentage decrease in the % attack success when the number of pigeons in the flock increases from 10 to 40.
2. Suggest an explanation for the effect of flock size on attack success shown in the table.
3. Some hawk species show cooperative hunting behaviour. Explain one advantage of this type of behaviour.

....

Q15: Match the following descriptions with the terms listed.

- Social signal used by the leader in a dominance hierarchy to assert authority
- Social signal used by low-ranking member of a social hierarchy to indicate acceptance of the dominant leader
- System of social organisation where the members are graded into a rank order
- Type of foraging behaviour employed by a group of predators resulting in mutual benefits

Terms: Cooperative hunting, Dominance hierarchy, Ritualised threat gesture, Subordinate response.

....

Q16: Ostriches are large birds which live on open plains in Africa. They divide their time between feeding on vegetation and raising their heads to look for predators.

The following graphs show the results of a study on the effect of group size on the behaviour of ostriches.

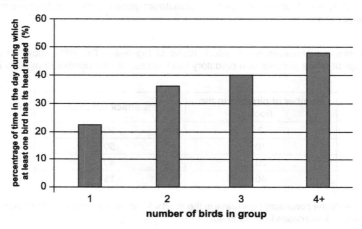

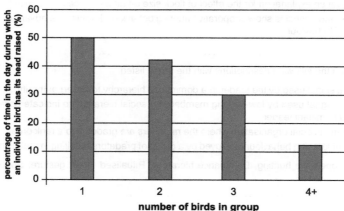

Which of the following is a valid conclusion from these results?

In larger groups, an individual ostrich spends:

a) less time with its head raised so the group is less likely to see predators.
b) less time with its head raised but the group is more likely to see predators.
c) more time with its head raised so the individual is more likely to see predators.
d) more time with its head raised but the group is less likely to see predators.

Q17: Which of the following examples of bird behaviour might be the result of social mechanisms for defence?

a) Great Tits with the widest stripe on their breast feed first when food is scarce.
b) Sooty Terns feed on larger fish than other species of tern which live in the same area.
c) Pelicans searching for food form a large circle round a shoal of fish, then dip their beaks into the water simultaneously.
d) Predatory gulls have difficulty picking out an individual puffin from a large flock.

Q18: The following list refers to pecking behaviour observed amongst six hens (P, Q, R, S, T and U).

P pecked U	P pecked T
R pecked T	S pecked P
S pecked U	T pecked Q
U pecked R	U pecked Q

Which bird was third in the pecking order?

Q19: What name is given to the type of social organisation that results in a rank order of individuals?

Q20: State two ways it is of advantage to the animals concerned.

Q21: Which of the following statements referring to advantages gained by hunting behaviour could be true of cooperative hunting?

1. Individuals gain more energy than from hunting alone.
2. Both dominant and subordinate animals benefit.
3. Much larger prey may be killed than by hunting alone.

a) 1 and 2 only
b) 1 and 3 only
c) 2 and 3 only
d) 1, 2 and 3

Q22: The honey bee is a social insect which lives in colonies. The queen is the only female in a colony that reproduces. Other females are workers which collect food, maintain the colony and care for the developing offspring.
Explain the advantage to the worker bees of caring for the offspring of the queen. *(2 marks)*

Q23: Colonial nesting birds - gulls and terns, for example - may provide formidable opposition to an invading predator such as a fox by mobbing it, even hitting the predator with their feet.

The graph below shows the pattern of predation on experimental eggs laid out near nests of colonial and solitary pairs of common gulls. 50 experimental eggs were placed out at the start of the experiment near each nest.

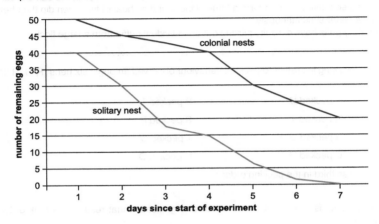

1. From the graph, calculate the percentage eggs left near solitary nests after 4 days from start of experiment.
2. Calculate the percentage decrease in eggs left near colonial nests after 7 days from start of experiment.
3. What conclusion can you reach from these results. Give a reason for your answer.

....

Q24: Termites and bees are examples of social insects. Give another example.

....

Q25: Complete the following sentence by picking a word from each bracket.

In social insects, (all/few) individuals breed and the offspring are raised by the (queen/workers). Most of the bees in a colony are (drones/queens/workers) that help to raise close relatives but do not themselves reproduce. This is an example of (social hierarchy/kin selection/reciprocal altruism).

....

Q26: Primates are social animals which often live in large groups. They display characteristic behaviours which enable group living.

Give one example of appeasement behaviour.

Unit 3 Topic 8

Components of biodiversity

Contents

8.1 Introduction 355
8.2 Genetic diversity 356
8.3 Species diversity 357
8.4 Ecosystem diversity 357
8.5 Learning points 358
8.6 Extended response question 359
8.7 End of topic test 359

> **Prerequisites**
>
> You should already know that:
>
> - biotic, abiotic and human influences are all factors that affect biodiversity in an ecosystem;
> - various factors can increase or decrease the biodiversity of an ecosystem.

> **Learning objective**
>
> By the end of this topic you should be able to:
>
> - understand that measurable components of biodiversity include:
>   - genetic diversity;
>   - species diversity;
>   - ecosystem diversity;
> - state that genetic diversity comprises the genetic variation represented by the number and frequency of all the alleles in a population;
> - understand that if one population dies out then the species may have lost some of its genetic diversity;
> - state that loss of genetic diversity may limit a species' ability to adapt to changing conditions;
> - know that species diversity comprises the number of different species in an ecosystem (species richness) and the proportion of each species in the ecosystem (the relative abundance);
> - explain how a community with a dominant species has lower species diversity than one with the same species richness but no particularly dominant species;
> - state that ecosystem diversity refers to the number of distinct ecosystems within a defined area.

TOPIC 8. COMPONENTS OF BIODIVERSITY

8.1 Introduction

Biodiversity refers to the variation of life on Earth. Studies indicate that the environments with the greatest biodiversity are:

- tropical rain forests;
- coral reefs;
- the deep sea;
- large tropical lakes.

Coral reef

Humans rely on the biodiversity of our planet to provide us with raw materials, foods, industrial chemicals and medicines. We also rely on biodiversity to provide ecosystem services such as pollination, purification of water, recycling of nutrients and natural pest control.

Loss of biodiversity reduces the availability of ecosystem services and other useful products. It also decreases the ability of species, communities, and ecosystems to adapt to changing environmental conditions. Biodiversity is nature's insurance policy against natural disasters. By maintaining high biodiversity it is likely that at least a few species will be able to adapt to any change in conditions which may occur.

Scientists measure biodiversity to gather information which will help them to conserve as many species and ecosystems as possible for the future. The measurable components of biodiversity include genetic diversity, species diversity and ecosystem diversity.

8.2 Genetic diversity

The Earth supports an extraordinary variety of interdependent life forms, upon which natural selection has been acting for millions of years. Each member of a species possesses hundreds or even thousands of genes. Since two or more alleles exist for most genes, the number of genetic combinations possible is enormous. The potential for genetic diversity amongst the members of a species is therefore immense.

In most natural populations, individuals vary slightly in their genetic makeup, which is why they do not all look or behave exactly alike. This is known as genetic diversity. Genetic diversity comprises the genetic variation represented by the number and frequency of all the alleles in a population. If one population dies out then the species may have lost some of its genetic diversity, and this may limit its ability to evolve successfully by adapting to changing environmental conditions.

The genetic diversity among individuals of a snail species is reflected in the variations in shell colour and banding patterns

Maintaining genetic variation among crop species, may be vital to the continuing success of crop development programmes. For example, in 1977 scientists discovered a previously unknown wild corn species, in South-central Mexico. This species happens to carry particularly useful genes, such as those for resistance to several viral diseases that affect domestic corn. Using these genes, scientists developed virus-resistant domestic corn varieties. Because corn is the third largest food crop on Earth, this discovery could prove critical to the global food supply.

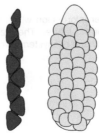

Ancestral form of corn and modern maize

8.3 Species diversity

The second level of **biodiversity** concerns species diversity upon which much public attention is focused. Species diversity comprises the number of different species in an ecosystem (the species richness) and the proportion of each species in the ecosystem (the relative abundance). Because species diversity is dependent upon both the species richness and their relative abundance, a community with a dominant species will have a lower species diversity than one with the same species richness but no particularly dominant species. This is shown in the table below. Both the grazed field and open meadow have the same species richness. However, the dominance of the grasses in the open meadow means the other species have a lower relative abundance, therefore reducing species diversity in this ecosystem.

Species	Grazed field	Open meadow
Grass	25	60
Buttercup	10	5
Broad-leaved plantain	15	10
Daisy	20	10
White clover	15	5
Dandelion	15	10

Relative abundance (%) of species in a grazed field and an open meadow

It is thought that there are about 5 to 10 million species on Earth at present. However, species are not constant unchanging units. Their number and kind are always changing. At any given moment, some species will be enjoying a stable relationship with the environment, some will be moving towards extinction and others will be forming new species.

8.4 Ecosystem diversity

Ecosystem diversity refers to the number of distinct ecosystems within a defined area. A great deal of attention has been paid to the level of species diversity in species-rich ecosystems such as tropical forests, but some scientists have argued that other relatively species-poor ecosystems are highly threatened and similarly need to be conserved.

In North America, prairie grassland once covered large expanses of the middle region of the country. Due to its fertility large areas have been used for agriculture and today less that 1% of the original tallgrass prairie remains. Some states such as Minnesota, Nebraska and Montana have put conservation programmes in place to conserve this ecosystem.

American prairie

8.5 Learning points

Summary

- The measurable components of biodiversity include genetic diversity, species diversity and ecosystem diversity.
- The genetic diversity comprises the genetic variation represented by the number and frequency of alleles in a population.
- If one population dies out then the species may have lost some of its genetic diversity. This loss may limit its ability to adapt to changing conditions.
- Species diversity comprises the number of different species in an ecosystem (the species richness) and the proportion of each species in the ecosystem (the relative abundance).
- A community with a dominant species has lower species diversity than one with the same species richness but no particular dominant species.
- Ecosystem diversity refers to the number of distinct ecosystems within a defined area.

TOPIC 8. COMPONENTS OF BIODIVERSITY

8.6 Extended response question

The activity which follows presents an extended response question similar to the style that you will encounter in the examination.

You should have a good understanding of biodiversity before attempting the question.

You should give your completed answer to your teacher or tutor for marking, or try to mark it yourself using the suggested marking scheme.

Extended response question: Biodiversity

Describe the measurable components of biodiversity. (*4 marks*)

8.7 End of topic test

End of Topic 9 test Go online

Q1: Give two components of biodiversity.

....

Q2: What do we call the number of different species in a habitat?

a) Species diversity
b) Relative abundance
c) Species richness

....

Q3: Which term refers to the variety of habitats?

a) Genetic
b) Ecosystem
c) Species

....

Q4: Which term refers to the variety in the gene pool of a species?

a) Genetic
b) Ecosystem
c) Species

....

Q5: Which term refers to the variety of living organisms found in different habitats?

a) Genetic
b) Ecosystem
c) Species

Q6: Which of the following ecosystems would tend to remain most stable?

	Relative state of the ecosystem	Predator-prey relationships
A	Simple	Only one prey species for each predator
B	Complex	Only one prey species for each predator
C	Simple	Many prey species for each predator
D	Complex	Many prey species for each predator

Q7: Which of the following best defines 'species diversity' of a habitat?

a) The maximum number of individuals which the resources of the island can support.
b) The number of different species in an ecosystem and the species richness.
c) The proportion of each species in an ecosystem and their relative abundance.
d) The number of different species in an ecosystem and the proportion of each species in the ecosystem.

Q8: The number and frequency of _____ in a population is a measure of genetic diversity.

Unit 3 Topic 9

Threats to biodiversity

Contents

9.1 Overexploitation 363
9.2 The impact of habitat loss 365
9.3 Introduced, naturalised and invasive species 368
9.4 Learning points 369
9.5 Extended response question 370
9.6 End of topic test 371

Prerequisites

You should already know that:

- biotic, abiotic and human influences are all factors that affect biodiversity in an ecosystem;
- various factors can increase or decrease the biodiversity of an ecosystem.

> **Learning objective**
>
> By the end of this topic you should be able to:
>
> - describe the problems associated with overexploitation of particular species and the impact on the genetic diversity of that species;
> - state that small populations can lose the genetic variation necessary to enable evolutionary response to environmental change;
> - understand that this phenomenon is known as the bottleneck effect;
> - understand that loss of genetic diversity can be critical for many species, as inbreeding results in poor reproductive rates;
> - state that some species have a naturally low genetic diversity in their population and yet remain viable;
> - state that habitat fragments typically support lower species richness than a large area of the same habitat;
> - understand that habitat fragments suffer from degradation at their edges and this may further reduce their size;
> - state that species adapted to the habitat edges (edge species) may invade the interior of the habitat at the expense of interior species;
> - state that, to remedy widespread habitat fragmentation, isolated fragments can be linked with habitat corridors;
> - understand that habitat corridors allow species to feed, mate and recolonise habitats after local extinctions;
> - give the meaning of the terms introduced, naturalised and invasive species and describe their impact on indigenous populations;
> - understand that invasive species may be free of the predators, parasites, pathogens and competitors that limit their population in their native habitat;
> - state that invasive species may prey on native species or outcompete them for resources.

9.1 Overexploitation

Humans make use of many different species to gain useful products such as raw materials and food; this is referred to as exploitation of natural resources. Overexploitation applies to a situation where individuals are being removed from a population at a greater rate than can be replaced by reproduction. If overexploitation continues, the species could become extinct from the area and the natural resource could be lost. A common example of overexploitation of a natural resource is overfishing.

During the 1970s more than 300,000 tonnes of cod were caught from the North Sea. This size of catch proved to be unsustainable and by 2006 it had fallen to less than 30,000 tonnes; a reduction of 90%. The population of cod in the North Sea had reduced in numbers to a level which could result in their extinction from the area. Thanks to conservation methods, such as reduced quotas and limits on the number of days fishing vessels can spend at sea, cod stocks are now on the increase, although they are still far from the level considered to be safe from collapse. It seems that in this instance, although overfishing reduced the numbers of cod in the North Sea the population is capable of recovering.

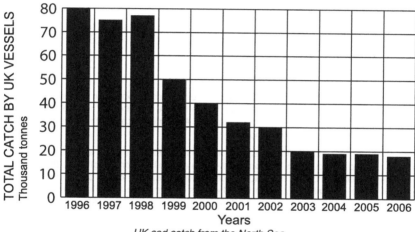

UK cod catch from the North Sea

The bottleneck effect

Loss of many individuals from a small species may result in a loss of the genetic variation necessary to enable evolutionary responses to environmental change. The **bottleneck** effect refers to a loss of large numbers of individuals within a species. This can occur by natural means such as forest fires or it can occur due to human activities such as overhunting. A bottleneck event results in a small population which may have lost some of its genetic variation. It is possible for the population to recover in numbers, however, this loss of **genetic diversity** effectively results in inbreeding which causes poor reproductive rates. The next diagram explains how a population bottleneck can result in loss of certain alleles (loss of genetic diversity) and a change in allele frequency.

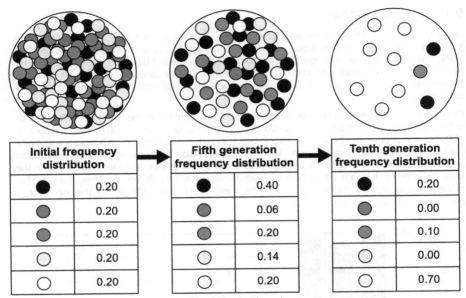
Population bottleneck

Cheetahs are an example of a species which have experienced a population bottleneck. Scientists have speculated that roughly 10,000 years ago, as the last ice age drew to a close, large numbers of cheetahs died out leaving very small populations in Asia and Africa. All the cheetahs now living are descended from this handful of individuals

Scientific research has shown that cheetahs from as far apart as East and Southern Africa - populations isolated by thousands of kilometres - were as similar to one another as 20 generations of deliberately inbred livestock or laboratory mice. When geneticists looked at the level of variation within genes of the cheetah, they found that cheetahs exhibit much lower levels of variation than other mammals. In most species, related individuals share about 80 per cent of the same genes. With cheetahs, this figure rises to approximately 99 percent. This genetic inbreeding in cheetahs has led to:

- low survival rates;
- greater susceptibility to disease;
- poor reproductive rates.

TOPIC 9. THREATS TO BIODIVERSITY

Cheetah (http://commons.wikimedia.org/wiki/File:Cheetah_0592.jpg by http://commons.wikimedia.org/wiki/User:Ltshears, licensed under http://creativecommons.org/licenses/by-sa/3.0/deed.en)

In another experiment, skin grafts were exchanged between members of two groups of cheetahs. In unrelated animals the survival time for skin grafts averages ten to twelve days. But in this experiment, all the grafts were accepted, though some were later slowly rejected. Several grafts persisted for at least 78 days, by which time they appeared to blend in with the recipient's own skin. This result indicated that the cheetahs' immune system did not recognise the tissue as being from another animal, and therefore failed to produce an immune reaction. Yet more evidence to show the **genetic uniformity** of populations of cheetahs.

Some species have a naturally low genetic diversity in their population and yet remain viable. Domesticated species often have low levels of genetic diversity. This is caused by the artificial selection of crops and animals for traits that humans find preferable.

9.2 The impact of habitat loss

An animal's habitat includes feeding sites, breeding grounds, burrowing sites and hunting areas. Human activities can split up such areas, causing animals to lose both their natural habitat and the ability to move between regions of an ecosystem. **Fragmentation** of habitats also means populations can be isolated and thus interbreeding may be prevented leading to a decrease in genetic diversity. Habitat fragments typically support lower species richness than a large area of the same habitat, thus reducing **biodiversity**. An additional issue is that habitat fragments suffer from **degradation** at their edges and this may further reduce their size.

Habitat fragmentation can have even greater effects than simply isolating populations. For example a forest ecosystem is not uniform across its area, within this ecosystem there is an interior which has different characteristics from the edges. Different organisms will be found within different regions of the forest. Fragmentation of ecosystems such as forests can result in changes to the ratio of edge habitat to interior habitat, as shown in the next diagram.

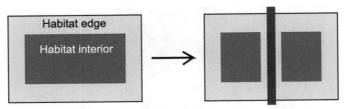

Habitat fragmentation

The centre of the forest is shaded by trees and has less wind and light than the forest edge, which is unprotected. Many forest-adapted species thus shy away from forest edges and prefer forest centres. Habitat fragmentation can result in an increased area of edge type habitat. Species adapted to the habitat edges may increase in number and invade the habitat core at the expense of interior species, reducing biodiversity.

Habitat fragmentation is a major problem across our planet. Roads, urbanisation and agriculture are among the main human activities which break up natural areas, often with disastrous implications for wildlife. A clear example which illustrates the importance of connectivity between fragmented habitats can be seen with the wood ants in the Scottish Caledonian forests. These forest-dwelling insects will not cross distances of more than 100 metres of open ground. Therefore, if wood ants are absent from an isolated area of forest, they will not be able to recolonise it, and the insect fauna of that woodland would be permanently depleted.

Habitat fragmentation due to human development is an ever-increasing threat to biodiversity, and habitat corridors are one possible solution. A habitat corridor is a strip of land that aids in the movement of species between disconnected areas of their natural habitat, allowing species to feed, mate and recolonise habitats after local extinctions. The main goal of implementing habitat corridors is to increase biodiversity. When areas of land are broken up by human interference, population numbers become unstable and many animal and plant species become endangered. By re-connecting the fragments, the population fluctuations can decrease dramatically.

The main benefits of habitat corridors are:

- colonisation: they allow animals to move and occupy new areas when food sources become scarce in their core habitat;
- migration: species can relocate seasonally without the need for human interference;
- interbreeding: animals can find new mates in neighbouring regions so that genetic diversity can increase within the population.

TOPIC 9. THREATS TO BIODIVERSITY

Ecoduct habitat corridor (http://en.wikipedia.org/wiki/File:Cerviduct.jpg by http://nl.wikipedia.org/wiki/Gebruiker:Henkmuller, licensed under http://creativecommons.org/licenses/by-sa/3.0/deed.en)

The impact of habitat loss: Questions Go online

Indian Tigers were formerly distributed evenly from Nepal into Bhutan, Northern India through Thailand, Cambodia and Malaysia. Nowadays, however, due to urbanisation and habitat destruction the tiger is only found in small populations as shown on the following map.

Present day distribution of Indian Tigers

© HERIOT-WATT UNIVERSITY

Q1: Give the term used to describe the process which has restricted the tiger to these eight areas.

....

Q2: Suggest one method which could be taken to avoid extinction of the tiger in these countries.

....

Q3: With reference to genetic diversity, explain how this method could improve the survival chances of the tiger. (*2 marks*)

9.3 Introduced, naturalised and invasive species

Introduced (non-native) species are those that humans have moved either intentionally or accidentally to new geographical locations. They are often called exotic species. Those that become established within wild communities are termed **naturalised** species. Most often the species are introduced for agricultural purposes, or as sources of timber, meat, or wool, and these species need humans for their continued survival. Some species have been introduced for **aesthetic** purposes. Others, such as plants, insects, or marine organisms, are unintentionally transported via the movement of cargo by ships or planes. Regardless of their method of introduction, some introduced species become **invasive**. Invasive species are naturalised species that spread rapidly and eliminate native species.

Invasive species often fare better in the new environment due to lack of predators, parasites, pathogens and competitors that limit their population in their native habitat. For example cane toads are native to Central and South America but were introduced to Australia as a form of biological control to kill crop pests. The toads quickly became invasive due to the lack of predators in the Australian ecosystem.

Cane toad (http://commons.wikimedia.org/wiki/File:Adult_Cane_toad.jpg by https://www.flickr.com/photos/briangratwicke/, licensed under http://creativecommons.org/licenses/by/2.0/deed.en)

Other invasive species may prey on native species or outcompete them for resources. For example grey squirrels were introduced to the UK in the late 1800s. Since then, they have caused a dramatic reduction in red squirrel numbers due to their ability to outcompete them for food and nesting sites.

TOPIC 9. THREATS TO BIODIVERSITY

Grey squirrels also carry a disease which kills red squirrels but does not affect the grey squirrels.

Grey squirrel

Red squirrel

Introduced, naturalised and invasive species: Question

Q4: Match the descriptions with the terms listed.

- Established within wild communities
- Moved by humans either intentionally or accidentally to new geographical locations
- Species indigenous to the location
- Spread and outcompeting native species for space and resources

Terms: introduced, invasive, native, naturalised.

9.4 Learning points

Summary

- Overexploitation has greatly reduced the number of organisms in some populations, for example some fish species in the North Sea.
- Some populations of particular species have been able to recover even after the population has been greatly reduced.
- After a large decrease in numbers, a small population may lose the genetic variation necessary for evolutionary responses to environmental changes.
- This reduction in genetic variation is known as the 'bottleneck effect'.
- A population bottleneck is an evolutionary event in which a significant percentage of a population or species is killed or otherwise prevented from reproducing.

> **Summary continued**
> - This loss of genetic diversity may be critical to some species, as inbreeding results in poor reproductive rates.
> - The clearing of habitats has led to habitat fragmentation.
> - Habitat fragments typically support lower species richness than a large area of the same habitat.
> - More isolated fragments and smaller fragments exhibit a lower species diversity.
> - Habitat fragments suffer from degradation at their edges.
> - Species adapted to habitat edges (edge species) may invade the habitat at the expense of interior species.
> - Isolated fragments can be linked with habitat corridors.
> - The corridors allow movement of animals between fragments increasing access to food and choice of mate. This may lead to recolonisation of small fragments after local extinctions.
> - Introduced (non-native) species are those that humans have moved either intentionally or accidentally to new geographic locations.
> - These may have an impact on indigenous (native) populations.
> - Those that become established within wild communities are termed naturalised species.
> - Invasive species are naturalised species that spread rapidly and eliminate native species.
> - Invasive species may well be free of the predators, parasites, pathogens and competitors that limit their population in their native habitat.
> - Invasive species may also prey on native species and outcompete them for resources.

9.5 Extended response question

The activity which follows presents an extended response question similar to the style that you will encounter in the examination.

You should have a good understanding of introduced species before attempting the question.

You should give your completed answer to your teacher or tutor for marking, or try to mark it yourself using the suggested marking scheme.

> **Extended response question: Introduced species**
>
> Discuss introduced species and their impact on indigenous (native) populations. *(6 marks)*

9.6 End of topic test

End of Topic 10 test   Go online

Q5: Which of the following correctly describes how the number of species present in a habitat fragment compare to the number of species present in the original habitat?

a) Larger
b) Same
c) Smaller

....

Q6: What name is given to the links which can be made between isolated habitat fragments?

....

Q7: Give one reason why invasive species are able to spread rapidly and eliminate native species.

....

Q8: What term is used to describe the situation whereby a significant percentage of a population or species is killed or otherwise prevented from reproducing?

....

Q9: Which of the following statements about genetic diversity are correct?

a) It is a measure of genetic differences within and between individuals, populations and species.
b) It is the variety of genetic material within a single species of organism that permits the organism to adapt to changes in the environment.
c) Genetic variation, required for natural selection, increases after a bottleneck event.
d) If there is enough genetic variation after a bottleneck event, the species can still recover but will lack genetic diversity.

....

Q10: The following table illustrates a small range of biodiversity of organisms found in a wood and what some of the living organisms do at different times of the year.

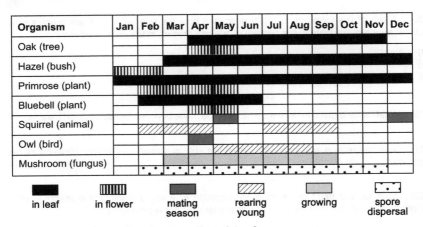

For how many months are there leaves on the oak tree?

....

Q11: What percentage of the year is the primrose plant in flower?

....

Q12: Bluebells live on the floor of the wood. Explain why it is an advantage to the bluebells to produce leaves in February rather than later in the year.

....

Q13: In terms of biodiversity, explain why this wood could be said to have a wide range of species diversity.

....

Q14: A population of squirrels in a wood were discovered to have a very low genetic diversity. Explain how this could have occurred.

....

Q15: Northern elephant seals experienced a population bottleneck caused by humans hunting them in the 1890s. At the end of the 19th century, hunting by man had reduced their population size to as few as 20 individuals. Their population now has risen to 30,000.

1. What effect has this had on their genetic diversity? Choose from:
   - increased;
   - decreased
   - stayed the same.
2. Give an example of a land animal which has a low genetic diversity.

TOPIC 9. THREATS TO BIODIVERSITY

Q16: The following table shows data obtained from an investigation into the biodiversity of species in a heathland food web.

	Species	Mean mass of organisms (g)	Population density (numbers m^{-2})
A	Cricket	0.20	4
B	Ladybird	0.04	30
C	Aphid	0.003	5240
D	Green lacewing	0.004	3225

Which row in the table shows correctly the species with the highest biomass per square metre?

Q17: The Sea Star *Pisaster Ochraceous* is a key predator found on rocks on the coast of certain areas of the USA, and it feeds on mussels and other invertebrates. The graph below shows the effect on the biodiversity of other species of removing and not removing *Pisaster* from rock pools in 1993.

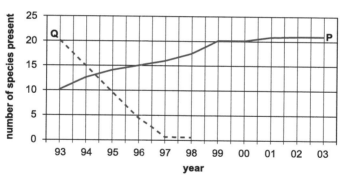

Which row in the following table correctly describes the results?

	Line P	Line Q	Role of *Pisaster*
A	With *Pisaster*	Without *Pisaster*	Increases species diversity
B	With *Pisaster*	Without *Pisaster*	Decreases species diversity
C	Without *Pisaster*	With *Pisaster*	Increases species diversity
D	Without *Pisaster*	With *Pisaster*	Decreases species diversity

Unit 3 Topic 10

End of unit test

End of Unit 3 test

Q1: Upon which process does all food production ultimately depend?

....

Q2: Why is food security becoming a global issue?

....

Q3: Which of the following factors will help to increase food production?

1. Breeding higher yielding cultivars.
2. Protecting crops from pests and diseases.
3. Changing land used for crops to livestock production.

a) 1 and 2 only
b) 1 and 3 only
c) 2 and 3 only
d) 1, 2 and 3

....

Q4: The diagram shows the fate of sunlight landing on a leaf.

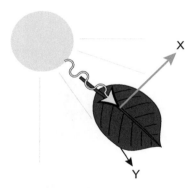

Which line in the following table correctly identifies the fate of sunlight represented by X and Y?

	X	Y
A	Transmission	Reflection
B	Absorption	Transmission
C	Reflection	Transmission
D	Reflection	Absorption

....

TOPIC 10. END OF UNIT TEST

Q5: The following diagram summarises the process of photosynthesis in a chloroplast.

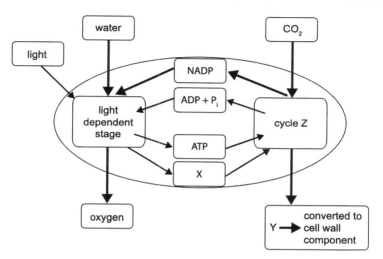

1. Name molecules X and Y.
2. Name cycle Z.
3. Name the enzyme responsible for fixing carbon dioxide into cycle Z.
4. Name the cell wall component referred to in the diagram.

....

Q6: Plant and animal breeding involves the manipulation of heredity to develop new and improved organisms to provide sustainable food sources. Name one characteristic which breeders may seek to improve in a crop organism.

....

Q7: Which field trial design decision would you take to eliminate bias when measuring treatment effects?

a) Number of replicates.
b) Randomisation of treatment.
c) Selection of treatments.

....

Q8: Which field trial design decision would you take to take account of the variability within a sample?

a) Number of replicates.
b) Randomisation of treatment.
c) Selection of treatments.

Q9: Which field trial design decision would you take to ensure fair comparison?

a) Number of replicates.
b) Randomisation of treatment.
c) Selection of treatments.

Q10: Selected plants or animals can be bred for several generations until the population breeds true to the desired type due to the elimination of heterozygotes. What is this process known as?

Q11: The following are some features of weed species. Which two describe the features of an annual plant weed?

a) Vegetative reproduction
b) Storage organs
c) High seed output
d) Rapid growth

Q12: Control of the whitefly with the parasitic wasp *Encarsia* is an example of:

a) natural control.
b) selective control.
c) biological control.
d) integrated control.

Q13: Name one problem which pesticides may cause to the environment.

Q14: The following list describes observed behaviour of pigs on a farm.

1. Lying in a position which does not allow suckling.
2. Repeated flicking of the head.
3. Frequent wounding of other pigs by biting.
4. Constantly bar biting.

Which of these behaviours indicate poor animal welfare?

a) 1, 2 and 3 only
b) 1, 3 and 4 only
c) 2, 3 and 4 only
d) 1, 2, 3 and 4

TOPIC 10. END OF UNIT TEST

Q15: Animals in captivity can show different behaviours from wild individuals of the same species. What name is given to a behaviour which involves unusual repetitive movement?

....

Q16: Which of the following lines best describes the effects of a parasitic relationship on the parasite and the host?

a) Benefits the parasite and benefits the host.
b) Benefits the parasite and harms the host.
c) Harms the parasite and benefits the host.
d) Harms the parasite and harms the host.

....

Q17: Like many animals, termites have microorganisms which live in their guts. The termites receive cellulose digesting enzymes from the microorganisms which allow them to use wood as a food source. The microorganisms are provided with a safe place to live. State the term used to describe this type of symbiotic relationship.

....

Q18: State one way in which parasites can be transmitted.

....

Q19: The following list shows benefits which an animal species can obtain from certain types of social behaviour.

1. Aggression between individuals is controlled.
2. Subordinate animals are more likely to gain an adequate food supply.
3. Experienced leadership is guaranteed.
4. Energy used by individuals to obtain food is reduced.

Which statements refer to social hierarchy?

a) 1, 2 and 3 only
b) 1, 2 and 4 only
c) 1, 3 and 4 only
d) 1, 2, 3 and 4

....

Q20: Other than termites, give an example of a social insect.

....

Q21: Altruistic behaviour is often observed between individuals which are closely related. What name is given to this form of altruism?

....

Q22: Primates, such as chimpanzees, often use appeasement behaviour to reduce unnecessary conflict within the group. Give one example of this type of behaviour.

....

© HERIOT-WATT UNIVERSITY

Q23: Give one feature of parental care in primates which allows complex social behaviour to be learned.

....

Q24: Which component of biodiversity is indicated by the number and frequency of alleles in a population?

....

Q25: The number of different species in a habitat is called:

a) species diversity.
b) species richness.
c) relative abundance.

....

Q26: If a dominant grass species invades an open meadow, what effect will this have on species diversity in the area?

a) It will increase.
b) It will decrease.
c) It will stay the same.

....

Q27: The following illustration shows two habitat fragments produced as a result of deforestation and the remaining forest.

Which habitat fragment is likely to have the greatest species diversity?

....

Q28: Suggest one measure which could be taken to link the isolated habitat fragments to the main forest.

....

TOPIC 10. END OF UNIT TEST

Q29: At a certain point in their life history, the numbers of cheetahs in the wild reduced drastically. As a result, populations of cheetah now show very little genetic variation. What name is given to this effect?

Read the following passage before answering the following three questions.

Many species of plant have been removed from their native habitat and brought to the UK, for example pink sorrel. In some cases this has been performed intentionally and in others, by accident. Some species find that they are able to establish themselves in the new environment and compete on an equal footing with the native species, for example the evening primrose. Other species brought to the UK spread rapidly and eliminate native species, for example rhododendron.

Q30: Which of the species is naturalised?

a) Evening primrose
b) Rhododendron
c) Pink sorrel

....

Q31: Which of the species is invasive?

a) Evening primrose
b) Rhododendron
c) Pink sorrel

....

Q32: Which of the species is introduced (non-native)?

a) Evening primrose
b) Rhododendron
c) Pink sorrel

A group of students performed an experiment to investigate the rate of photosynthesis in lupin and foxglove plants.

Five leaf discs were cut from each plant and suspended in a solution that provided carbon dioxide in syringes. Air was removed from the discs which caused them to sink and the apparatus was placed in a dark room.

The discs were illuminated at a low light intensity by a lamp from above. The more quickly the leaf discs floated, the greater their rate of photosynthesis. The following diagram shows the positions of the leaf discs after fifteen minutes.

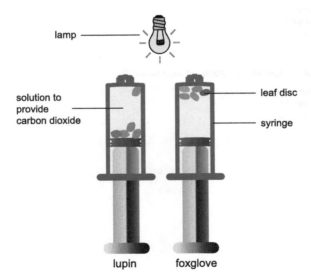

Q33: Why were five leaf discs used in each tube?
....

Q34: Name two variables which must be kept constant when setting up this experiment.
....

Q35: Why did the leaves which photosynthesised float?

....

Q36: Foxgloves are shade plants which grow well on the forest floor in the shade of large trees. Explain how the results show that foxgloves are well adapted as shade plants.

....

Q37: In a related investigation, the rate of photosynthesis in a foxglove was measured at different light intensities. The results from this experiment are shown in the following table.

Light intensity (kilolux)	Rate of photosynthesis (units)
10	4
20	20
30	55
40	86
50	90
60	90

Draw a line graph to display the results of this investigation.

....

Q38: Calculate the percentage increase in the rate of photosynthesis when light intensity increases from 20 kilolux to 30 kilolux.

....

Q39: Shade plants, such as foxgloves, have adaptations which allow them to use green light for photosynthesis. Suggest an adaptation which would allow a plant to absorb a wider range of wavelengths of light for photosynthesis.

Glossary

Absorbed
: the light which is taken into a plant leaf by pigments

Activation energy
: the minimum energy required by reactants to allow reaction to occur

Active site
: the region of an enzyme molecule where the enzyme acts on the substrate

Aesthetic
: branch of philosophy dealing with the nature of beauty and art

Aestivation
: dormancy in response to high temperature or drought

Agriculture
: the process of producing feed and other desirable products by the cultivation of certain plants and the raising of domesticated animals

Allele
: one of the different forms of a gene

Allele frequency
: the prevalence of alternative versions of genes

Anabolic
: a reaction which requires energy and builds up molecules

Animal welfare
: physical and psychological well-being of animals. The term animal welfare can also mean human concern for animal welfare. Welfare is measured by indicators including behaviour, physiology, longevity, and reproduction

Annual weed
: plant which grows, flowers, set seeds and dies within the space of one year

Anticodon
: a triplet of exposed bases on a tRNA molecule

Antiparallel
: running in an opposite direction

Apical meristems
: growing points (regions of mitosis) found at the tips of plant stems or roots allowing increase in length

Archaea
: a group of single-celled microorganisms

GLOSSARY

Artificial selection
  intentional breeding controlled by humans for particular traits or characteristics

ATP
  adenosine triphosphate, coenzyme used as an energy carrier in the cells of all known organisms

ATP synthase
  an enzyme which produces ATP

Biodiversity
  degree of variation of life forms within a given species, ecosystem, biome, or an entire planet

Bioinformatics
  a process which combines computer science and statistical analysis to study genomes

Biological catalysts
  catalysts made of protein that are only found in living cells

Biomass
  the total mass of living matter within a given unit of environmental area

Blastocyst
  an embryo that has developed for 5 to 6 days after fertilisation

Bottleneck
  an evolutionary event in which a significant percentage of a population or species is killed or otherwise prevented from reproducing

Calorimeter
  a piece of equipment used to measure heat generation from an organism to allow metabolic rate to be calculate

Calvin Cycle
  a series of biochemical reactions that takes place in the chloroplast and does not require light

Carnivore
  animal which eats meat and which derives its energy requirements from a diet consisting mainly or exclusively of animal tissue whether through predation or scavenging

Catabolic
  a reaction which releases energy and breaks down molecules

Chlorophyll
  the green pigment which is found in almost all plants and green algae. It absorbs light which is essential for photosynthesis

Chloroplast
  the photosynthetic unit of a plant cell, containing all the chlorophyll

Citric acid cycle
> the second stage of respiration, where acetyl CoA and oxaloacetate join to form citrate and a series of reactions which return citrate to oxaloacetate

Codon
> a triplet of exposed bases on a length of mRNA

Colonial
> relating to a colony

Competition
> an interaction or struggle between organisms or species for a resource such as food, territory or mates, in which the fitness or numbers of one is reduced by the presence of another

Competitive inhibition
> competitive inhibition of enzyme activity occurs when an inhibitor, resembling the structure of the substrate, binds to the active site of the enzyme and blocks the binding of the substrate

Cooperative hunting
> animals such as lions hunt as a group to increase their chances of successfully killing prey

Cultivar
> plant or group of plants selected for a particular characteristic

Daily torpor
> a period of reduced activity in organisms with high metabolic rates

Degradation
> process by which ecosystems or habitats are broken down or fragmented

Dehydrogenase
> an enzyme which removes hydrogen ions and electrons from substrates

Deletion
> removal of a length of DNA from a chromosome

Deletion mutation
> loss of a section of DNA or a number of nucleotides

Differentiation
> the process by which cells or tissues undergo a change towards a more specialised function

DNA ligase
> an enzyme that facilitates the process by which fragments of DNA are joined together

DNA polymerase
> an enzyme that synthesises DNA strands from individual nucleotides

Dormancy
> a condition of biological rest or inactivity characterised by cessation of growth or development and the suspension of many metabolic processes

GLOSSARY

Double helix
the double helical shape of a DNA molecule

Duplication
repetition of a series of nucleotides within a chromosome

Effector
cells, muscles or glands which perform responses to stimuli

Electron transport chain
the final stage of respiration where high energy electrons and hydrogen ions are used to synthesise ATP

Eukaryote
an organism which possesses a membrane-bound nucleus

Exons
the parts of the initial mRNA which are used to code for proteins

Feedback inhibition
regulation of enzyme activity where the first enzyme of a metabolic pathway is inhibited by the reversible binding of the final product of the pathway

Fermentation
a type of respiration which takes place in the absence of oxygen

Food security
the ability of human populations to access food of sufficient quality and quantity

Fragmentation
habitat fragmentation describes the emergence of discontinuities (fragmentation) in an organism's environment (habitat), causing population fragmentation

Fungicide
a chemical compound or biological organism used to kill or inhibit fungi or fungal spores

G-3-P
glycerate-3-phosphate, which is a substance found in the carbon fixation stage (Calvin Cycle) of photosynthesis

Gene pool
complete set of unique alleles in a species or population

Genetic diversity
comprises the genetic variation represented by the number and frequency of alleles in a population

Genetics
the branch of biology that deals with heredity, especially the mechanisms of hereditary transmission and the variation of inherited traits among similar or related organisms

Genetic uniformity
when the genes or alleles of a population are similar and show little variation

Genome
the entirety of an organism's hereditary information

Genomics
the science of interpreting genes; the study of an organism's genome using information systems, databases and computerised research tools

Genotype
a statement of an organism's alleles for a particular characteristic usually given as symbols - a pea plant could have the genotype CC if it were homozygous for pink petal colour or the genotype Cc if it were heterozygous for pink colour

Glycolysis
the first stage of respiration where glucose is broken down into pyruvate

Herbicide
a chemical compound used to kill unwanted plants

Herbivore
an organism adapted to eat plant-based foods, such as deer, cows and sheep

Heterotrophic
an organism which gains energy by consuming other organisms

Heterozygotes
having two different alleles for a characteristic - a pea plant heterozygous for petal colour has two different petal colour alleles Cc

Hibernation
an inactive state resembling deep sleep in which certain animals living in cold climates pass the winter

Hierarchy
an organisation arranged in a graded order with member(s) at the top who are dominant over subordinate individuals

Homozygotes
having two identical alleles for a characteristic - a pea plant homozygous for petal colour has two identical petal colour alleles, both pink (CC) or both white (cc)

Hypothalamus
part of the brain which monitors and regulates temperature

Inbreeding
the reproduction from mating two genetically related parents

Inbreeding depression
the reduced fitness in a given population as the result of breeding of related individuals

GLOSSARY

Induced fit model
 a model of an enzyme-substrate reaction that causes a conformational change in the active site of the enzyme that allows the substrate to fit perfectly

Induced pluripotent stem cells
 somatic (adult) cells reprogrammed to enter an embryonic stem cell-like state

Insecticide
 a chemical compound used to kill insects

Insertion mutation
 the addition of an extra nucleotide

Introns
 the parts of the initial mRNA which are removed before translation

Invasive
 introduced species (also called "non-indigenous" or "non-native") that adversely affect the habitats they invade economically, environmentally, and/or ecologically

Inversion
 the inversion (reversal) of a section of DNA within a chromosome

Lagging strand
 the strand of DNA that grows in the direction opposite to the movement of the growing fork; it is replicated in fragments

Leading strand
 the strand of DNA that is being replicated continuously

Legume
 a plant which is able to fix atmospheric nitrogen to synthesis amino acids which can then be built up to plant proteins. This is due to the symbiotic relationship with bacteria in the root nodules of these plants

Ligase
 an enzyme which joins fragments of DNA together

Light reaction
 the photosynthetic process in which solar energy is harvested and transferred into the chemical bonds of ATP; can occur only in light

Livestock
 one or more domesticated animal raised in an agricultural setting to produce commodities such as food, fibre and labour. The term does not usually involve farmed fish

Meristem
 a growing point in a plant, i.e. a place where mitosis produces new cells

Messenger RNA
    (mRNA) is synthesised from a DNA template, resulting in the transfer of genetic information from the DNA molecule to the messenger RNA

Metabolites
    the intermediates and products of metabolic reactions that take place in organisms

Migration
    a process which avoids metabolic adversity by expending energy to relocate to a more suitable environment

Misdirected behaviour
    abnormal behaviour which the animal directs at another object, animal or human

Mitochondria
    a structure in the cell responsible for producing energy

Mitochondrion
    a structure in the cell responsible for producing energy

Mitosis
    nuclear division

Mollusc
    a large group of invertebrate organisms including slugs

Monoculture
    the agricultural practice of producing or growing one single crop over a wide area. It is widely used in modern agriculture and its implementation has allowed for large harvests of crops from minimal labour

NAD
    a co-enzyme which easily attaches to hydrogen ions, but releases them when they are required

NADP
    nicotinamide adenine dinucleotide phosphate is a coenzyme which is used to carry hydrogen (NADPH) to chemical reactions which require a reducing agent

Naturalised
    any process by which a non-native organism spreads into the wild and its reproduction is sufficient to maintain its population

Natural selection
    the survival of the fittest, whereby only individuals with the most suitable genetic constitution for any set of circumstances pass their genes on

Negative feedback
    homeostasis; the process by which an increase in one factor causes a decrease in another factor, thereby maintaining equilibrium around a set point (norm)

GLOSSARY

Nematode
>organisms which belong to the group known as the roundworms and can be found in almost every ecological system

Non-competitive inhibition
>a molecule binds to part of the enzyme away from the active site, and causes a conformational change in the active site of the enzyme, thereby inhibiting the binding of the appropriate substrate molecule

Nutrient medium
>a mixture of nutrients (including carbon and nitrogen sources) required for growth

Outbreeding
>the practice of introducing unrelated genetic material into a breeding line

Pecking order
>a natural hierarchy in a group of birds, such as domestic fowl

Perennial
>a plant which lives for more than two years

Persistent
>chemical compounds which do not break down or degrade easily in the environment

Placatory
>leading to a reduction in tension, to pacify or appease

Plasmid
>a circular, self-replicating DNA molecule that carries only a few genes

Plastome
>the genetic material that is found in plastids in plant cells (for example in the chloroplast). It composes part of the entire genome of photosynthetic organisms

Pluripotent stem cells
>these are stem cells, with the potential to make any differentiated cell in the body

Polyculture
>agriculture using multiple crops in the same space

Polyploid
>organisms which contain more than two set of chromsomes

Predator
>an organism that feeds on another organism

Primer
>a strand of nucleic acid that serves as a starting point for DNA

Producer
an organism which uses light energy (green plants) or chemical energy (some bacteria) to manufacture the organic compounds it needs as nutrients from simple inorganic compounds obtained from its environment

Productivity
the rate of generation of biomass in an ecosystem. It is usually expressed in units of mass per unit surface (or volume) per unit time, for instance grams per square metre per day

Prokaryote
an organism which lacks a membrane-bound nucleus

Receptor
cells which monitor changes in environment

Reflected light
light which is bounced off a leaf and does not get absorbed and is not available for photosynthesis

Respirometer
a piece of equipment used to measure the rate of respiration

Restriction endonuclease
an enzyme that cuts specific target sequences of DNA

Ribosomal RNA
(rRNA) is the RNA that is a permanent structural part of a ribosome

Ribosomes
structures found in the cytoplasm where protein synthesis occurs

RNA splicing
a process which removes introns from a primary mRNA transcript

RuBisCO
ribulose-1,5-bisphosphate carboxylase oxygenase, is an enzyme involved in carbon fixation (Calvin Cycle) that catalyzes the first major step of carbon fixation, a process by which the atoms of atmospheric carbon dioxide are made available to organisms in the form of energy-rich molecules such as carbohydrates

RuBP
ribulose-1,5-bisphosphate is an organic substance that is involved in photosynthesis

Sanitation
the removal of crop residues and unharvestable (perhaps pest-infected) plants that might harbour pest insects from outside the crop area

Selectively permeable
a property of a membrane which means that substances do not freely pass through it; the membrane allows the passage of certain small molecules, but excludes many other molecules

GLOSSARY

Self-renewal
: a property displayed by stem cells which allows them to divide to produce more stem cells

Speciation
: the formation of a new species

Species
: group of organisms which can interbreed to produce fertile, viable offspring

Stereotypic behaviour
: repetitive or ritualistic movement, posture, or utterance, found in animals with welfare problems

Substitution mutation
: the replacement of one nucleotide by another

Sustainable
: a pattern of resource use that aims to meet human needs while preserving the environment so that these needs can be met not only in the present, but also for generations to come

Symbiosis
: close and often long-term interaction between different biological species

Synthesis
: the building up of complex molecules from simpler ones

Transcription
: the production of mRNA from a DNA template

Transfer RNA
: (tRNA) is a short strand of RNA that is twisted on itself to expose three bases, and which carries a specific amino acid to a ribosome

Translation
: the sequencing of amino acids at ribosomes, based on the sequence of nucleotides in mRNA

Translocation
: transposition of a length of DNA onto another chromosome

Transmitted light
: transmitted light is light which passes right though the leaf

Trophic level
: the position or stage an organism occupies in a food chain. Trophic levels can be represented by numbers, starting at level 1 (or A) for plants

Vasoconstriction
: contraction in diameter of a blood vessel, thus reducing blood flow

Vasodilation
: enlargement in diameter of a blood vessel, thus increasing blood flow

Weed
    a plant that is considered to be a nuisance, and normally applied to unwanted plants in human-controlled settings, especially farm fields and gardens

Wild-type
    describes the phenotype of the typical form of a species as it occurs in nature

Answers to questions and activities for Unit 1

Topic 1: Structure and organisation of DNA

The structure of DNA: Questions (page 8)

Q1:

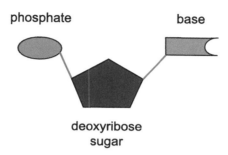

Q2:

- The nucleotide guanine pairs with **cytosine**.
- The nucleotide thymine pairs with **adenine**.
- The nucleotide cytosine pairs with **guanine**.
- The nucleotide adenine pairs with **thymine**.

Q3: Double helix

Q4: Hydrogen

Q5: Phosphate, deoxyribose sugar, base

The organisation of DNA in prokaryotes and eukaryotes: Questions (page 10)

Q6: Prokaryotes

Q7: Mitochondria / Chloroplasts

Q8: Plasmid

End of Topic 1 test (page 13)

Q9: Genome

Q10: TACCTGAAATCCA

Q11: 20%

Q12: 4. C, A, B

Q13: Deoxyribose

Q14: Hydrogen bonds

Q15: Sugar-phosphate

Q16:

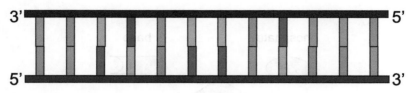

Q17: Mitochondria / Chloroplasts

Q18: Protein / Histones

Topic 2: Replication of DNA

Extended response question: DNA structure and replication (page 23)

Suggested marking scheme

Each line represents a point worth one mark. The concept may be expressed in other words. Words which are bracketed are not essential. Alternative answers are separated by a solidus (/); if both such answers are given, only a single mark is allocated. In checking the answer, the number of the point being allocated a mark should be written on the answer paper. A maximum of eight marks can be gained.

DNA structure *(maximum of 4 marks)*:

1. DNA can be linear or circular.
2. DNA is made up of two strands of nucleotides.
3. A nucleotide consists of a deoxyribose sugar, a phosphate group and base.
4. Nucleotides are linked together by strong chemical bonds between the deoxyribose sugar of one nucleotide and the phosphate group of another.
5. The bases of DNA always pair up: adenine with thymine, and guanine with cytosine.
6. There are hydrogen bonds between bases.
7. DNA takes the shape of a double helix...
8. ...with antiparallel strands / deoxyribose and phosphate at 3' and 5' end of each strand.

DNA replication *(maximum of 4 marks)*:

i. During replication, the hydrogen bonds between the bases in the DNA molecule break and the strands unwind.
ii. A primer binds to each strand of DNA.
iii. DNA polymerase replicates a strand of DNA from free DNA nucleotides.
iv. DNA polymerase adds complementary nucleotides to the 3' end (of the lead chain) OR in one direction.
v. One strand is replicated continuously.
vi. The other (lag) strand is replicated in fragments.
vii. Fragments are joined by ligase.

End of Topic 2 test (page 25)

Q1: DNA polymerase

Q2: Primer

Q3: DNA ligase

Q4: 2. b, a, d, c - as per:
- The hydrogen bonds between DNA strands break.
- Base pairing occurs between free nucleotides and each of the DNA strands.
- Nucleotides are bonded together by DNA polymerase.
- The DNA molecules coil up to form double helices.

Q5: a) The enzyme is relatively stable at high temperatures.

Q6: 3. d, c, b, f, a, e - as per:
- Temperature of the reaction adjusted to 92-98°C.
- The DNA strands separate.
- Temperature of the reaction adjusted to 50-65°C.
- Annealing of the primers to the single-stranded DNA.
- Temperature of the reaction adjusted to 70-80°C.
- Synthesis of DNA by the enzyme DNA polymerase.

Q7: Extension occurs at 70-80°C.

Q8: PCR leads to an **exponential** amplification of desired DNA sequences.

Q9: b) 8

Topic 3: Gene expression

The structure and functions of RNA: Questions (page 31)

Q1:

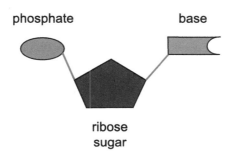

Q2:

	RNA	DNA
Structure	not a double helix	double helix
Preferred form	single-stranded	double-stranded
Number of types	>1	1
Present in	the cytoplasm and the nucleus	the nucleus
Bases	adenine, uracil, guanine, cytosine	adenine, thymine, guanine, cytosine

Q3: Phosphate, ribose sugar, base

Q4: (A) adenine, (U) uracil, (G) guanine, (C) cytosine

Q5: mRNA, tRNA, rRNA

Q6: mRNA copies the code from the DNA molecule and carries it out to the ribosomes where the proteins are synthesised.
tRNAs are found in the cytoplasm, attaching to specific amino acids and bringing them to the ribosomes where the amino acids are joined together.

Q7: c) both DNA and RNA

Transcription: Questions (page 36)

Q8: RNA polymerase

Q9: Nucleus

Q10: Hydrogen bonds

Q11: Primary transcript / Pre-mRNA

Q12: RNA splicing

Translation: Questions (page 39)

Q13:

1. The double-stranded DNA unwinds, hydrogen bonds in the DNA break and the DNA strands separate
2. An RNA nucleotide binds to a complementary nucleotide on one of the DNA strands
3. Hydrogen bonds form between the bases on the RNA and the DNA nucleotides
4. The RNA nucleotides are linked together to form messenger RNA (mRNA)
5. When synthesis of the mRNA is completed, the mRNA separates from the DNA
6. The mRNA leaves the nucleus and enters the cytoplasm
7. A ribosome attaches to the mRNA. Two transfer RNA (tRNA) molecules are also contained within the ribosome
8. Hydrogen bonds are formed between the first codon of the mRNA and the complementary anticodon on a tRNA
9. The second tRNA binds to the mRNA
10. A peptide bond forms between the amino acids carried by the tRNA molecules
11. The first tRNA leaves the ribosome, and another tRNA enters and base-pairs with the mRNA
12. A second peptide bond is then formed. The process continues, with the ribosome moving along the mRNA
13. As each mRNA codon is exposed, incoming tRNA pairs with it and polypeptide synthesis continues until completed

Q14: Anticodon

Q15: Cytoplasm

Q16: Ribosome

Q17: Peptide bond

Q18: They cause translation to stop / They are stop codons

Q19:

DNA	C	A	C	A	G	T	G	T	T	T	G	T	C	C	G
mRNA	G	U	G	U	C	A	C	A	A	A	C	A	G	G	C
protein		val			ser			gln			thr			gly	

Protein structure and function: Questions (page 43)

Q20:

Description	Diagram
Chain of amino acids linked by strong peptide bonds	d)
Polypeptide structure determined by weak hydrogen bonds	c)
Strong bonds form between special groups of amino acids	a)
More than one polypeptide makes up the final structure	b)

Extended response question: Protein synthesis (page 45)

Suggested marking scheme

Each line represents a point worth one mark. The concept may be expressed in other words. Words which are bracketed are not essential. Alternative answers are separated by a solidus (/); if both such answers are given, only a single mark is allocated. In checking the answer, the number of the point being allocated a mark should be written on the answer paper. A maximum of seven marks can be gained.

1. mRNA carries information / code (for proteins) from the nucleus / from DNA.
2. mRNA attaches to ribosome.
3. Three bases on mRNA is a codon.
4. tRNA transports amino acids to ribosome.
5. tRNA transports specific amino acids.
6. Three bases on tRNA is an anticodon.
7. Codons match / pair with their anticodons.
8. Joins / adds correct amino acid onto growing protein/polypeptide.
9. Sequence of bases / codons on mRNA gives sequence of amino acids.

End of Topic 3 test (page 45)

Q21: 3. A, C and B - as per:

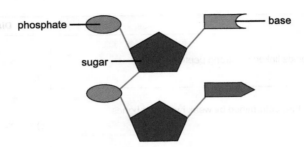

Q22: c) It is single-stranded; it has ribose in its backbone; it contains the base uracil

Q23:

a) CGUAAGUAACGU
b) Transcription
c) RNA polymerase

Q24:

The mRNA produced after transcription is called the **primary transcript**; the **introns** are removed, leaving only the **exons** in the final **mature transcript**.

Q25: RNA splicing

Q26: Translation

Q27: CGVASL - as per:

tRNA molecules

C	G	V	A	S	L
ACG	CCA	CAG	CGA	AGG	GAC
UGC	GGU	GUC	GCU	UCC	CUG

mRNA molecule

Q28: Anticodon

Q29: Alternative splicing

Q30: Hydrogen

Topic 4: Differentiation in multicellular organisms

Stem cells: Questions (page 56)

Q1: Stem cells have the ability to go through numerous cycles of cell division while maintaining the undifferentiated state.

Q2: Stem cells are undifferentiated cells; this allows them to undergo the process of cell differentiation and gives them the ability to form other types of body cells.

Q3:

- A muscle cell: movement;
- a red blood cell: transport of oxygen;
- a nerve cell: carries impulses.

If you have a different answer, check it with your teacher as it may still be correct.

Embryonic stem cells: Question (page 58)

Q4: The stages of the process of using hESCs to form specialised cells are as follows:

1. Early human embryo Blastocyst
2. Embryo stem cell removed
3. Stem cell cultured in the laboratory
4. Formation of specialised cells: nerve cell, muscle cell, gut cells
5. Undifferentiated stem cells cultured in different culture conditions
6. Formation of mass of undifferentiated stem cells

Extended response question: Stem cells (page 63)

Suggested marking scheme

Each line represents a point worth one mark. The concept may be expressed in other words. Words which are bracketed are not essential. Alternative answers are separated by a solidus (/); if both such answers are given, only a single mark is allocated. In checking the answer, the number of the point being allocated a mark should be written on the answer paper. A maximum of six marks can be gained.

Embryonic stem cells *(maximum of 2 marks)*:

1. Found in developing embryo (blastocyst).
2. Have the capacity to become all cell types.
3. Can be grown relatively easily in culture.

Tissue stem cells *(maximum of 4 marks)*:

  i. Found in body tissues.
  ii. Are more differentiated than embryonic stem cells.
  iii. Can only differentiated into a narrow range of cell types.
  iv. Are rare in mature tissues.
  v. Give rise to a limited range of cell types.
  vi. Develop into cell types that are closely related to the tissue in which they are found.

End of Topic 4 test (page 65)

Q5: c) It contains specialised cells that differentiate

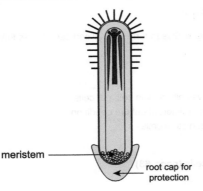

Q6: **D**, as per

Q7: An unspecialised cell.

Q8: They self-replicate and differentiate.

Q9: Bone marrow, skin epidermis, brain, blood, other suitable organs. *(pick any two)*
If you have a different answer, check it with your teacher as it may still be correct.

Q10: To replace differentiated cells.

Q11:

- Skin grown from stem cells to treat burn victims.
- Bone marrow transplant of stem cells to treat leukaemia.
- Production of replacement organs such as a windpipe for transplants.
- Testing new drugs using stem cells.

Q12: It involves the destruction of embryos.

Topic 5: Structure of the genome

The genome: Question (page 70)

Q1:

Process	Description
Transcription	DNA copied to RNA
Splicing	Introns removed from pre-mRNA
Translating	Exons pass to ribosome where polypeptides are assembled

End of Topic 5 test (page 72)

Q2: The **genome** of an organism is its **hereditary** information encoded in **DNA**.

Q3: c) genes

Q4: False

Q5: mRNA, tRNA, rRNA, RNA fragments *(pick any two)*

Q6: False

Q7: False

Topic 6: Mutations

Gene mutations: Questions (page 79)

Q1:

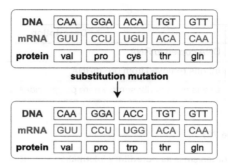

Q2:

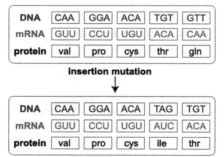

Q3:

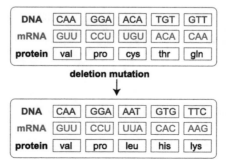

ANSWERS: UNIT 1 TOPIC 6

Q4:

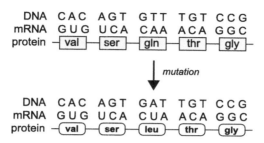

Q5:

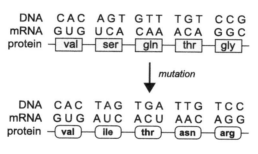

Q6:

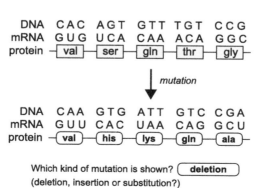

Effects of gene mutations on amino acid sequences: Questions (page 84)

Q7: 4 amino acids have been changed; a deletion mutation is a **frameshift** mutation.

Q8: 4 amino acids have been changed; an insertion mutation is a **frameshift** mutation.

Q9: 1 amino acid has been changed; a substitution mutation is a **point** mutation.

Q10: The deletion of three nucleotides may 'cancel out' the frame shift mutation if the deletion corresponds to a base triplet that is transcribed into a mRNA codon. The pattern of base triplets in the DNA appearing after the mutation will be unaffected. There will be a loss of one amino acid from the protein, but the other amino acids will remain the same. The answer is illustrated by the example given:

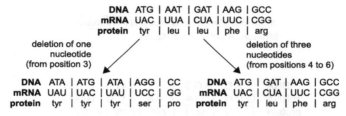

Q11: A mutation may introduce a 'stop' codon (UAA, UAG or UGA) into the encoded mRNA. Synthesis of the protein will cease when a stop codon is encountered in the mRNA during translation, and a shorter protein will be produced as a result. Here is an example of how this might occur:

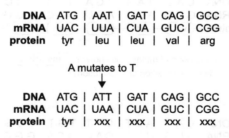

Differences between gene and chromosome mutation: Questions (page 87)

Q12: Chromosome mutation - deletion

Q13: Gene mutation - deletion

Q14: Gene mutation - insertion

Q15: Gene mutation - substitution

Q16: Chromosome mutation - duplication

Q17: Chromosome mutation - translocation

ANSWERS: UNIT 1 TOPIC 6

Extended response question: Gene mutations (page 90)

Suggested marking scheme

Each line represents a point worth one mark. The concept may be expressed in other words. Words which are bracketed are not essential. Alternative answers are separated by a solidus (/); if both such answers are given, only a single mark is allocated. In checking the answer, the number of the point being allocated a mark should be written on the answer paper. A maximum of six marks can be gained.

Gene mutations *(maximum of 3 marks)*:

1. A gene mutation is the replacement or altering of a single nucleotide within a DNA sequence.
2. A substitution mutation means that one nucleotide is substituted for another.
3. An insertion mutation means that one or more nucleotides are inserted into the DNA.
4. A deletion mutation means that one or more nucleotides are removed from the DNA.

Consequences *(maximum of 5 marks)*:

i. The effect of a mutation will depend on its type / location.
ii. Substitution mutations can result in a missense mutation where a single incorrect amino acid is inserted into a protein.
iii. It can also result in a nonsense mutation which results in the code for an amino acid being changed to a stop codon.
iv. Splice-site mutations result in some introns being retained and/or some exons not being included in the mature transcript.
v. Insertions / deletions of 1 or 2 nucleotides mean that all the bases downstream are moved up or down from their place; this means the reading frame is altered.
vi. This type of mutation is known as a frame-shift mutation.

End of Topic 6 test (page 91)

Q18: d) they are random, infrequent occurrences.

Q19: d) Translocation

Q20: a) Deletion

Q21: c) Insertion, deletion, substitution - as per:

Normal sequence		N	N	**C**	**A**	**C**	**G**	**T**	**A**	**A**	**C**	**G**	**T**	N	N
Insertion	A	N	N	**C**	**A**	**C**	**G**	**T**	**A**	**A**	**C**	**C**	**G**	**T**	N
Deletion	B	N	N	**C**	**A**	**G**	**T**	**A**	**A**	**C**	**G**	**T**	N	N
Substitution	C	N	N	**C**	**A**	**C**	**G**	**A**	**A**	**A**	**C**	**G**	**T**	N	N

Q22: c) Insertion or deletion

Q23:

1. Nonsense
2. Missense

Topic 7: Evolution

Natural selection: Questions (page 99)

Q1:

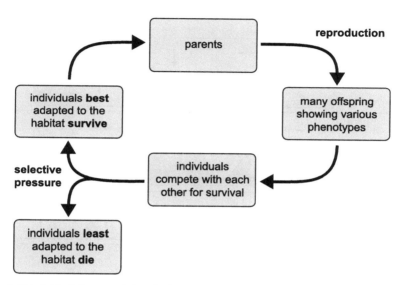

Q2: c) All individuals in a population display the same phenotypes.

Q3: b) The reproductive success it experiences during its lifetime.

Species: Questions (page 101)

Q4: A population is a group of individuals that belong to the same species and can interbreed with each other.

Q5: The gene pool is the sum of all the different alleles contained within a population. The allele frequency is the abundance of any given allele in a population.

End of Topic 7 test (page 106)

Q6: Evolution can be described as a change in a **species** over time, and is driven by **natural selection**.

Q7: Inheritance can be described as the passing of **genes / genetic material** between generations.

Q8: The exchange of plasmids between bacteria is an example of **horizontal** inheritance.

Q9:
   i. 12000
   ii. 400%

Q10:

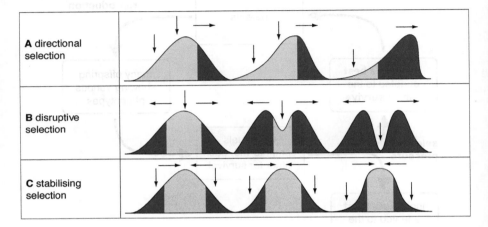

A directional selection	
B disruptive selection	
C stabilising selection	

Q11:
   i. Stabilising selection
   ii. Disruptive selection
   iii. Directional selection

Q12: False

Q13: Sympatric speciation uses behavioural or ecological barriers.

Topic 8: Genomics

Phylogenetics: Question (page 113)

Q1: All life forms are now described as belonging to one of three **domains**. This is largely based on a comparison of **DNA**. The three main groups are **bacteria**, **archaea** and **eukaryota**. Phylogenetic clocks need to be calibrated by using **fossil records**.

End of Topic 8 test (page 120)

Q2: Determining the order of nucleotide bases is known as **sequencing**.

Q3: Computer and statistical analyses.

Q4: To show the relationship between organisms.

Q5: Fossil evidence

Q6: The **genome** is the sum total of an organism's DNA.

Q7: No

Q8: The genes are the same / very similar.

Q9: Screening for genetic defects, better targeted treatment, lower costs of treatment.

Q10: Pharmacogenetics

Topic 9: End of unit test
End of Unit 1 test (page 122)

Q1: Hydrogen bond

Q2: 15%

Q3: Deoxyribose sugar

Q4: X - phosphate; Y - deoxyribose sugar; Z - base.

Q5: Primer

Q6: Nucleotides / ATP.

Q7: DNA polymerase

Q8: The **hydrogen** bonds of DNA are separated by **heating** during PCR.

Q9: c) Primers unwind double helix.

Q10: b) 8

Q11: Transcription

Q12: Nucleus

Q13: Translation

Q14: Ribosome

Q15: RNA polymerase

Q16: Choose an answer from:
- the introns / non-coding regions of genes are removed;
- due to RNA splicing;
- the mature mRNA only contains exons / coding regions of genes.

Q17: UAC

Q18: Peptide bonds

Q19: Hydrogen bonds

Q20: Cells that are undifferentiated / unspecialised.

ANSWERS: UNIT 1 TOPIC 9

Q21: Choose an answer from:
- Is it safer than using the drug directly on volunteers?
- Is it right to use embryos to extract stem cells?
- Is it right to deprive sufferers of potential treatment?
- Is it right to use stem cells rather than animals?

Q22: Tissue stem cells are more differentiated than embryonic stem cells.

Q23: Meristems are regions of **unspecialised / undifferentiated** cells in plants that are capable of cell **division**.

Q24: A region of the DNA molecule which codes for a protein.

Q25: tRNA / rRNA / RNA fragments

Q26: b) Substitution

Q27: Missense

Q28: Substitution

Q29: 2. Duplication

Q30: Bacteria can on occasions pass genetic material between themselves. This may be a section of DNA called a **plasmid**.

Q31: This is representative of **horizontal** genetic transfer.

Q32: c) Prokaryotes and eukaryotes

Q33:
- **Stabilising** selection is where the average phenotype is most successful for a particular habitat.
- **Disruptive** selection is characterised by the extreme versions of a phenotype being selected.
- **Directional** selection is characterised by the selection of one extreme phenotype at the exclusion of all others.

Q34: A group of organisms which can interbreed to produce fertile offspring.

Q35: Geographical

Q36: They cannot interbreed.

Q37: Sympatric

Q38: DNA sequencing

Q39: Chimps are **more** closely related to gorillas than orangutans.
The common ancestor of chimps and gorillas is **more** recent than the common ancestor of gorillas and orangutans.

© HERIOT-WATT UNIVERSITY

Q40: Bioinformatics

Q41: Comparison of genomes reveals that many genes are highly **conserved** across different organisms.

Q42: Pharmacogenetics

Q43: Control

Q44: To allow a comparison with those which were exposed to gamma radiation.

Q45: Choose an answer from:
- spacing of the chickpeas;
- volume of water;
- length of time exposed to gamma radiation.
- ph;
- oxygen concentration

Q46: As the gamma radiation dose increases from 0 to 800 Gy, the percentage germination decreases from 99% to 39%. As the radiation increases further to 900, Gy percentage germination increases to 53%.

Q47: 68%

Q48: Choose an answer from:
- gamma radiation does not affect germination of desi chickpeas;
- gamma radiation affects kabuli chickpeas more than desi chickpeas.

Answers to questions and activities for Unit 2

Topic 1: Metabolic pathways

Enzyme properties: Question (page 140)

Q1:

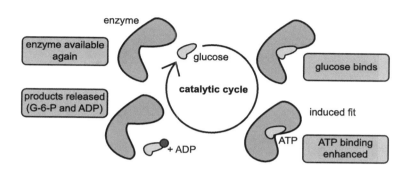

Competitive inhibition: Questions (page 143)

Q2: The active site is a small area of the enzyme, approximately 20 amino acids in length, where (depending on the reaction catalysed by the enzyme) one or more substrate molecules are bound, and where the reaction occurs.

Q3: Both the competitive inhibitor and the substrate are structurally similar, so that the inhibitor is able to bind to the active site.

Q4: The addition of more substrate to the experimental reaction increases the chances of the substrate molecule colliding with the enzyme and thereby overcoming competitive inhibition.

Feedback inhibition: Question (page 145)

Q5: It is energetically efficient as it avoids the excessive (and wasteful) production of the intermediates of a pathway.

Answers from page 146.

Q6:

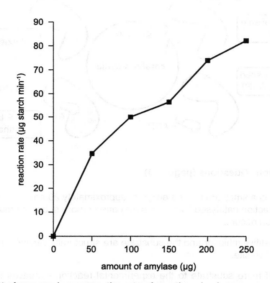

Q7: As the amount of enzyme increases, the rate of reaction also increases.

Q8: The graph indicates that the rate of reaction is not directly proportional to the amount of amylase. If it were, the result would be a straight line.

Q9: These tubes are experimental controls. Tube 6 shows that when there is no enzyme added the reaction does not occur. Tube 7 shows that when no substrate is added the reaction does not occur.

Q10: The reaction rate should continue to increase as the amount of amylase is increased. In practice, the increase in the reaction rate will become less and less as the amount of amylase is increased. This is because there is a finite amount of substrate present and once that has been converted to product (maltose) the enzyme becomes ineffective.

Extended response question: The control of the enzyme activity by inhibition (page 148)

Suggested marking scheme

Each line represents a point worth one mark. The concept may be expressed in other words. Words which are bracketed are not essential. Alternative answers are separated by a solidus (/); if both

ANSWERS: UNIT 2 TOPIC 1

such answers are given, only a single mark is allocated. In checking the answer, the number of the point being allocated a mark should be written on the answer paper. A maximum of eight marks can be gained.

1. Competitive inhibition...
2. ... is where an inhibitor competes with the substrate for the active site of an enzyme.
3. Competitive inhibition can be reversed by an increase in the concentration of the substrate.
4. Non-competitive inhibition...
5. ... is where an inhibitor binds to an enzyme away from the active site.
6. The shape of the active site is altered and the activity of the enzyme is reduced.
7. This type of inhibition may or may not be reversible.
8. Feedback inhibition...
9. ... is used in the control of metabolic pathways.
10. The end-product of the pathway inhibits the activity of the first enzyme in the pathway.

End of Topic 1 test (page 148)

Q11: A mutation in gene 4 would result in enzyme E4 not being produced (or not functioning properly). This means that metabolite D would build up.

Q12: All the reactions that take place in an organism.

Q13: Reactions which release energy are said to be **catabolic** and reactions which require energy are described as **anabolic**.

Q14: Enzymes

Q15: They allow molecules to pass across the plasma membrane.

Q16: d) Competitive inhibitor, enzyme, non-competitive inhibitor.

Q17: d) When the substrate binds to the active site, the shape of the active site is changed.

Q18: Enzymes **lower** the activation energy and release products with a **low** affinity for the active site.

Q19: a) Enzyme 1

Q20: a) 0 to 5 minutes from the start of the reaction.

Q21: c) a non-competitive inhibitor

Q22: **Competitive** inhibitors decrease the activity of an enzyme by binding to the active site.

Q23: b) Feedback inhibition

Topic 2: Cellular respiration

The role of ATP: Questions (page 156)

Q1:

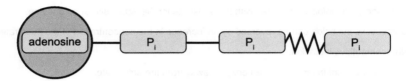

Q2: A chemical reaction that uses energy to build up complex molecules from simple molecules.

Q3: The synthesis of starch from glucose units / protein synthesis from amino acids.

Q4: An energy-releasing reaction that breaks down complex molecules to form simpler ones.

Q5: The breakdown of glycogen to glucose / the breakdown of glucose to carbon dioxide and water.

Glycolysis: Questions (page 157)

Q6:

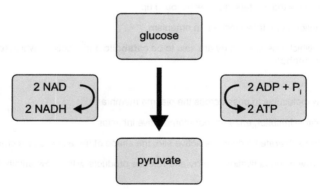

Q7: Cytoplasm

Q8: b) 2

Q9: No

Q10: Pyruvate

ANSWERS: UNIT 2 TOPIC 2

Citric acid cycle: Questions (page 159)

Q11:

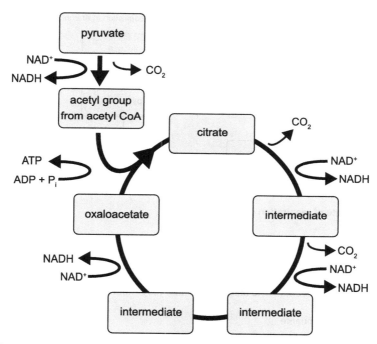

Q12: Citrate

Q13: c) Mitochondrion

Q14: Aerobic

Anaerobic respiration: Questions (page 161)

Q15:

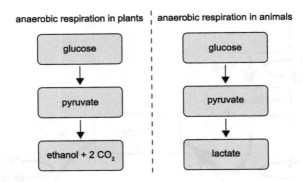

Measuring the rate of respiration (page 162)

Q16: c) 0.3 ml

Q17: d) 0.6

Extended response question: The stages of cellular respiration (page 164)

Suggested marking scheme

Each line represents a point worth one mark. The concept may be expressed in other words. Words which are bracketed are not essential. Alternative answers are separated by a solidus (/); if both such answers are given, only a single mark is allocated. In checking the answer, the number of the point being allocated a mark should be written on the answer paper. A maximum of eight marks can be gained.

A) Glycolysis *(maximum of 4 marks)*:
1. Glycolysis is an anaerobic process / does not require oxygen.
2. It occurs in the cytoplasm.
3. Glucose is split into pyruvate.
4. The hydrogen ions and electrons released during glycolysis bind to the coenzyme NAD to form NADH.
5. Dehydrogenase enzymes remove H ions and electrons, which are passed to the coenzyme NAD.
6. Two molecules of ATP are required to convert glucose to pyruvate; this is an energy investment phase.
7. Four molecules of ATP are produced by glycolysis (a net gain of two ATP molecules).

B) Citric acid cycle *(maximum of 4 marks)*:
  i. The citric acid cycle is an aerobic process that occurs in the matrix of the mitochondrion.
  ii. Pyruvate is converted to acetyl CoA.
  iii. Acetyl group binds to oxaloacetate to form citrate.
  iv. Citrate is converted, through a series of enzyme-catalysed reactions, back into oxaloacetate.
  v. In the process, both carbon (in the form of carbon dioxide) and hydrogen ions with electrons are released.
  vi. Hydrogen ions and electrons become bound to NAD to form NADH.
  vii. Dehydrogenase enzymes remove H ions and electrons, which are passed to coenzymes NAD. *(cannot be used again here if used in the glycolysis answer)*

End of Topic 3 test (page 164)

Q18: The production of ATP from ADP and P_i is called **phosphorylation**.

Q19: The breakdown of ATP releases **energy**, some of which is used in the synthesis of complex molecules.

Q20: b) It is a catabolic reaction that does not require oxygen.

Q21: d) Pyruvate

Q22: d) In the cytoplasm.

Q23: The citric acid cycle occurs in the **matrix** of the mitochondrion.

Q24: During the citric acid cycle **oxaloacetate** combines with an acetyl group to form **citrate**, this is gradually turned back into **oxaloacetate** by a series of **enzyme** controlled reactions.

Q25: c) Oxygen

Q26: ATP synthase

Q27: NAD

Topic 3: Metabolic rate
End of Topic 4 test (page 172)

Q1: Respirometer

Q2: a) Heart rate

Q3: 1642.5

Q4: A fish heart has **two** chambers.
An amphibian heart has **three** chambers.

Q5: It improves the efficiency of oxygen delivery / separating oxygenated and deoxygenated blood.

Q6: A mammal has a **double** circulatory system whereas fish have a **single** circulatory system.

ANSWERS: UNIT 2 TOPIC 4

Topic 4: Metabolism in conformers and regulators

Regulating body temperature: Question (page 184)

Q1:

Decrease in body temperature	Increase in body temperature
inactive sweat glands	active sweat glands
increase in metabolic rate	decrease in metabolic rate
hair erector muscles relaxed	hair erector muscles contracted
vasoconstriction	vasodilation

Q2: Hypothalamus

Q3: d) Shivering and vasoconstriction

Q4: Both sweating and flushing are mechanisms used by the body to cool itself down. When you sweat, the evaporation of the water from your skin cools you down. The skin flushes because the blood vessels have become vasodilated. This increases the amount of heat lost from the body by enabling more blood to flow to the skin surface.

Extended response question: Internal body temperature regulation in mammals (page 186)

Suggested marking scheme

Each line represents a point worth one mark. The concept may be expressed in other words. Words which are bracketed are not essential. Alternative answers are separated by a solidus (/); if both such answers are given, only a single mark is allocated. In checking the answer, the number of the point being allocated a mark should be written on the answer paper. A maximum of seven marks can be gained.

1. The hypothalamus is the temperature monitoring centre.

2. Information is communicated by electrical impulses through nerves to the effectors.

3. Effectors bring about corrective responses to return temperature to normal.

4. Vasodilation - increased blood flow to the skin increases heat loss. OR Vasoconstriction - decreased blood flow to skin decreases heat loss.

5. Increased temperature / body too hot leads to (increase in) sweat production OR converse.

6. Body heat used to evaporate water in the sweat, cooling the skin.

7. Decrease in temperature causes hair erector muscles to raise / erect hair traps (warm) air OR forms insulating layer.

8. Decrease in temperature causes muscle contraction / shivering which generates heat/raises body temperature.

9. Temperature regulation involves / is an example of negative feedback.

© HERIOT-WATT UNIVERSITY

10. When body temperature increases, metabolic rate is decreased so less heat produced. OR When body temperature decreases, metabolic rate is increased so more heat produced.

End of Topic 5 test (page 188)

Q5: a) Light intensity

Q6: Enzyme activity works around an optimum range, above it becomes denatured, below it is inactive.

Q7: The internal environment of **conformers** is dependent upon the external environment. **Regulators** control their internal environment.

Q8: Advantage: low metabolic requirements.
Disadvantage: narrow niche range.

Q9: Hypothalamus

Q10: Vasoconstriction

Q11: a) Decreased metabolic rate, and d) Vasodilation

Q12: Nerves *or* electrical impulses

Topic 5: Maintaining metabolism

Dormancy: Questions (page 193)

Q1:

Term	Definition
Hibernation	Period of long-term inactivity in animals
Aestivation	Dormancy in response to hot, dry conditions
Daily torpor	Period of short-term inactivity in animals

End of Topic 6 test (page 196)

Q2: b) Consequential dormancy

Q3: a) aestivation

Q4: c) hibernation

Q5: a) consequential dormancy.

Q6: c) both predictive and consequential dormancy.

Q7: 95

Q8: Successful migration depends on a combination of **innate** and **learned** behaviours.

Q9: *Any from:*
- capture and release;
- direct observation;
- radio tracking;
- tagging.

Topic 6: Environmental control of metabolism
Bacterial culture growth phases: Questions (page 210)

Q1: b) Q
Q2: c) R
Q3: a) P
Q4: d) S

Extended response question: The patterns of growth shown by microbes (page 213)

Suggested marking scheme
Each line represents a point worth one mark. The concept may be expressed in other words. Words which are bracketed are not essential. Alternative answers are separated by a solidus (/); if both such answers are given, only a single mark is allocated. In checking the answer, the number of the point being allocated a mark should be written on the answer paper. A maximum of six marks can be gained.

1. Growth of microorganisms can be followed on a growth curve.
2. Growth phases are lag, log, stationary and death.
3. During the lag phase microorganisms adjust to the conditions of the culture...
4. ... by producing enzymes that metabolise the available substrates.
5. During the log phase the rate of growth is at its highest.
6. During the stationary phase the culture medium becomes depleted...
7. ... and secondary metabolites are produced.
8. During the death phase a lack of substrate and the toxic accumulation of metabolites cause death of cells.
9. Primary metabolites are produced in the early phases.
10. Secondary metabolites are produced later.

ANSWERS: UNIT 2 TOPIC 6

End of Topic 7 test (page 215)

Q5: Archaea, bacteria and eukaryotes.

Q6: *Any two from:*

- gaseous environment;
- light;
- ph;
- temperature.

Q7: D

Q8:

1. Lag phase - organisms acclimatising to their environment.
2. Log phase - organisms growing exponentially.
3. Stationary phase - organisms likely to be producing secondary metabolites.
4. Death phase - decline of population due to exhaustion of nutrients and build-up of toxins.

Q9: d) 3.0

Q10: b) 4000

Topic 7: Genetic control of metabolism

Bovine somatotrophin: Questions (page 226)

Q1: Cloning vectors are DNA molecules that can transfer foreign DNA into a bacterial cell and replicate there.

Q2: A plasmid containing foreign DNA.

Q3: b) Restriction endonuclease enzymes

Q4: b) Restriction endonuclease enzymes

Q5: a) Ligases

Q6:

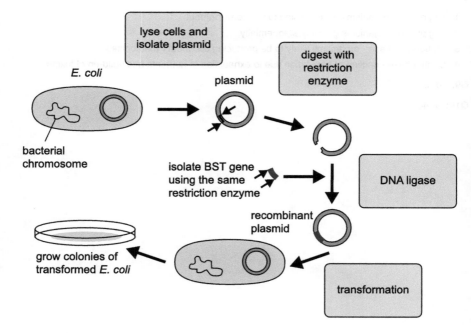

ANSWERS: UNIT 2 TOPIC 7

End of Topic 8 test (page 230)

Q7: False

Q8: Chemical and radiation.

Q9: Vector

Q10: Restriction endonuclease

Q11: Ligase

Q12: *Any one from*:
- marker genes;
- restriction site;
- genes for self-replication;
- regulatory sequences.

Q13: It may be folded incorrectly.

Q14: Recombinant yeast cell.

Topic 8: End of unit test
End of Unit 2 test (page 232)

Q1: enzymes

Q2: anabolic

Q3: catabolic

Q4: Substrate

Q5: When a substrate binds the active site on the enzyme changes shape.

Q6: Competitive inhibitors bind at the active site, non-competitive inhibitors bind away from the active site.

Q7: B

Q8: C

Q9: ATP synthase

Q10: Dehydrogenase

Q11: Glycolysis

Q12: Cytoplasm

Q13: NAD

Q14: During the citric acid cycle, **oxaloacetate** combines with an acetyl group to form **citrate**; this is gradually turned back into **oxaloacetate** by a series of **enzyme**-controlled reactions.

Q15: a) Calorie intake

Q16: c) Fish, amphibian, mammal

Q17: Salinity / pH / temperature

Q18: Hypothalamus

Q19: Negative feedback

Q20: To maintain optimum temperature for enzymes.

Q21: A **conformer** has an internal environment which is dependent on the external environment. They have **low** metabolic costs and inhabit a **narrow / small** range of ecological niches.

Q22: A **regulator** uses energy to control its internal environment. They have **high** metabolic costs and inhabit a **wide / large** range of ecological niches.

Q23: Some animals avoid extreme or hostile environments by **migration**.

Q24: Successful migration depends on a combination of **innate** and **learned** behaviours.

ANSWERS: UNIT 2 TOPIC 8

Q25: Consequential

Q26: Predictive

Q27: Archaea, bacteria, eukaryotes

Q28: As a result of their adaptability, microorganisms are found in a **wide / large** range of ecological niches.

Q29: *Any two from*:
- gaseous environment;
- light;
- pH;
- temperature.

Q30:
A) Lag.
B) Log / exponential.
C) Stationary.
D) Death.

Q31: Any from:
- mutagenesis;
- recombinant DNA.

Q32: Horizontal

Q33: Restriction endonuclease

Q34: Ligase

Q35: When working with **microorganisms** in a lab, as a safety mechanism, genes are often introduced that prevent the survival of the **microorganism** in an **external** environment.

Q36: Using a recombinant yeast cell.

Q37: To prevent cross-contamination.

Q38: Any from:
- mass of sugar added to the nutrient broth;
- volume of nutrient broth;
- type of nutrient broth;
- pH of nutrient broth;
- strain of *E. coli*.

Q39:

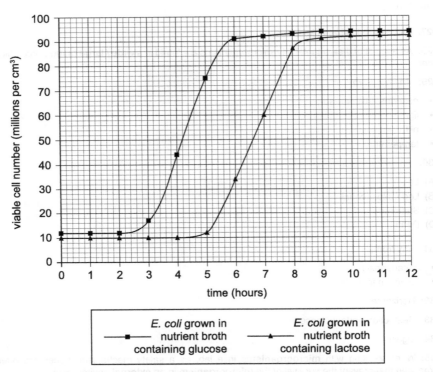

(1 mark for plotting the points correctly and connecting them with a ruler; 1 mark for filling in the key correctly)

Q40: The total cell numbers includes dead cells.

Q41: 2.5 hours / 150 minutes

Q42: *Any two from:*
- The lag phase is shorter when glucose is used as a respiratory substrate / the lag phase is longer when lactose is used as a respiratory substrate.
- The log phase begins earlier when glucose is used as a respiratory substrate / the log phase begins later when lactose is used as a respiratory substrate.
- The log phase lasts longer when glucose is used as a respiratory substrate / the log phase is shorter when lactose is used as a respiratory substrate.
- The stationary phase begins earlier when glucose is used as a respiratory substrate / the stationary phase begins later when lactose is used as a respiratory substrate.

Answers to questions and activities for Unit 3

Topic 1: Food supply

Agricultural production - Food production and photosynthesis: Questions (page 250)

Q1:

Raw materials for photosynthesis	Essential requirements	Products of photosynthesis
Water	Light	Sugar
Carbon dioxide	Chlorophyll	Oxygen

Q2: 32.4

Q3: 142.5

Q4: *Any two from:*

- Climate change may affect increased wheat production.
- New cultivars/crop plants may not yield increases in wheat production.
- Lack of available, high-quality agricultural land may restrict increase in wheat production.

Agricultural production - Trophic levels: Questions (page 253)

Q5: 33

Q6: 4

Q7: Energy transfer is inefficient because energy is lost while moving from one trophic level to another in the following ways:

- not the entire organism is consumed or digested - parts such as woody stems, bones, and scales are not eaten, and some materials such as cellulose cannot be digested;
- energy is used up by organisms in each trophic level for movement;
- energy is used in respiration and is released from the body of the organism as heat;
- energy becomes lost in excretion.

Q8:

- Energy released as respiration is used for movement and other life processes, and is eventually lost as heat to the surroundings.
- Energy is lost in waste materials, such as faeces.

Q9: There is more energy available in plants to human food chain than in plants to meat to human food chain as it is a shorter food chain with subsequently less energy loss. This would mean there would be more energy in plant food available to feed more people using plants-based diet than a meat-based diet.

End of Topic 1 test (page 257)

Q10: d) 0.25

Q11: b) producers.

Q12: c) Increased susceptibility to disease

Q13: a) trophic level A: producers.

Q14: c) 40,000 kJ $m^{-2} year^{-1}$

Q15: c) 1 kJ

Q16: Photosynthesis

Q17: To ensure food security OR because the human population is increasing

Topic 2: Plant growth and productivity

Photosynthetic pigments: Thin layer chromatography (page 264)

Q1: a) 0.39

Q2: c) 0.48

Q3: d) 0.64

Action spectrum: An experiment to determine an action spectrum for photosynthesis (page 268)

Q4: All four wavelengths contribute to photosynthesis, but the blue and red ends of the visible light spectrum are the major contributors to photosynthesis. The image shows a plot of the number of oxygen bubbles produced against the wavelength of light. This illustrates an action spectrum for photosynthesis.

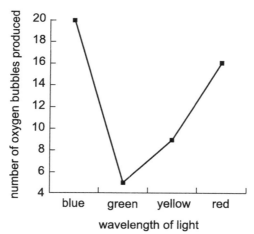

Q5: The rate of photosynthesis is significantly greater at the blue and red ends of the spectrum because chlorophyll can directly absorb light at these wavelengths.

Q6: The plant must be allowed to equilibrate to its new surroundings.

The light-dependent stage (page 270)

Q7:

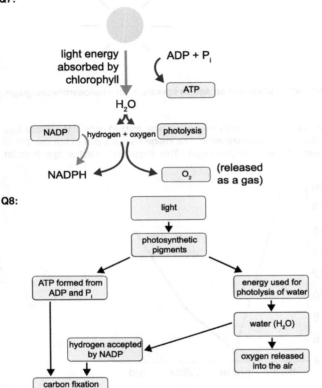

Q8:

The second stage of photosynthesis: The carbon fixation stage (page 271)

Q9:

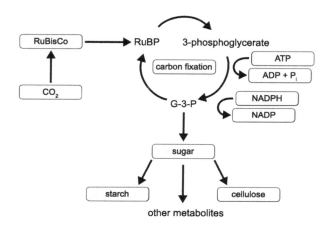

The second stage of photosynthesis: Questions (page 272)

Q10: X: Oxygen and Y: Glucose/carbohydrate

Q11: Photolysis of water

Q12: Calvin Cycle OR carbon fixation

Q13: Cellulose

Q14: *Any two from*:
- ATP
- NADPH
- hydrogen

Extended response question: Plant growth and productivity (page 274)

Suggested marking scheme

Each line represents a point worth one mark. The concept may be expressed in other words. Words which are bracketed are not essential. Alternative answers are separated by a solidus (/); if both such answers are given, only a single mark is allocated. In checking the answer, the number of the point being allocated a mark should be written on the answer paper. A maximum of sixteen marks can be gained.

A) Light and photosynthetic pigments in photosynthesis *(maximum of 8 marks)*:
1. Plants reflect, transmit and absorb light.
2. Only a small amount of the absorbed light energy is used in photosynthesis.
3. Photosynthetic pigments absorb light energy.
4. Chlorophyll a and chlorophyll b are the main photosynthetic pigments.
5. They absorb light in the red and blue range of the visible spectrum.
6. The carotenoids are accessory pigments.
7. They absorb light from other regions in the visible spectrum.
8. The accessory pigments pass the energy they absorb onto the chlorophyll.
9. The wavelengths of light that are absorbed by a pigment are called its absorption spectrum.
10. The wavelengths of light actually used by a pigment in photosynthesis are called its action spectrum.
11. The absorption spectrum of chlorophyll is closely related to the rate of photosynthesis.

B) Light-dependent stage of photosynthesis and the Calvin Cycle *(maximum of 8 marks)*:

Any four of the following for *4 marks*:

i. Absorbed energy excites electrons in the pigment molecule to raise them to high-energy levels.

ii. Transfer of these high-energy electrons through electron transport chain releases energy.

iii. This energy is used to generate ATP from ADP and P_i (inorganic phosphate).

iv. The enzyme ATP synthase is required for this process.

v. The light energy is used to split water molecules into oxygen and hydrogen.

vi. The hydrogen combines with the co-enzyme NADP forming NADPH.

vii. The oxygen is released from the leaf as a by-product of the reaction.

Any four of the following for *4 marks*:

I. The ATP and NADPH from the light dependent stage are transferred to the ~~Calvin Cycle~~ **carbon fixation stage**.

II. The enzyme RuBisCO fixes carbon dioxide from the atmosphere

III. by attaching it to RuBP.

IV. The 3-phosphoglycerate produced is phosphorylated by ATP and combined with the hydrogen from NADPH to form G-3-P.

V. G-3-P sugar may be synthesised into starch, cellulose or other metabolites.

VI. G-3-P is used to regenerate RuBP to continue the cycle.

VII. Major biological molecules in plants such as proteins, fats, carbohydrates and nucleic acids are derived from the photosynthetic process.

End of Topic 2 test (page 275)

Q15: d) use light of different wavelengths for photosynthesis.

Q16:

1. Transmitted
2. Chloroplasts

Q17:

1. Photolysis
2. NADP
3. Oxygen

Q18: The accessory pigments extend the wavelengths of light which can be absorbed by the plant.

Q19:

1. CO_2
2. RuBisCO

Q20: ATP

© HERIOT-WATT UNIVERSITY

Q21: C

Q22:

Term	Description
Oxygen	Product of the photolysis of water which is required for aerobic respiration.
NADP	Compound which accepts hydrogen during the photolysis of water.
Water	Raw material which becomes split into oxygen and hydrogen during photolysis of water.
ADP + P_i	Components of a high-energy compound.
Photolysis	Breakdown of water during the light- dependent stage of photosynthesis.
Hydrogen	Product of photolysis of water which becomes attached to NADP.
Chlorophyll	Green pigment which traps light energy.
Light dependent reaction	First stage in photosynthesis in which light energy is converted to chemical energy.

Q23:

Term	Description
Chloroplast	Structure found in a leaf where photosynthesis takes place.
NADPH	Hydrogen acceptor needed for the fixation of carbon in carbohydrates.
RuBP	Carbon compound which acts as a carbon dioxide acceptor.
G-3-P	First stable compound formed in the carbon fixation stage (Calvin Cycle) after carbon dioxide combines with its acceptor molecule.
Carbon dioxide	Raw material which supplies carbon atoms to be fixed into carbohydrates.
ATP	High-energy compound used to phosphorylate the intermediate compound in carbon fixation (Calvin Cycle).
Carbon fixation	Second stage in photosynthesis which is also known as the carbon fixation stage (Calvin Cycle).
RuBisCO	The enzyme which fixes carbon dioxide by attaching it to RuBP.

Q24: Marking Scheme: Axis with appropriate scales plus labels with units (all of table headers): (*1 mark*). Points plotted accurately: (*1 mark*).

ANSWERS: UNIT 3 TOPIC 2

1.
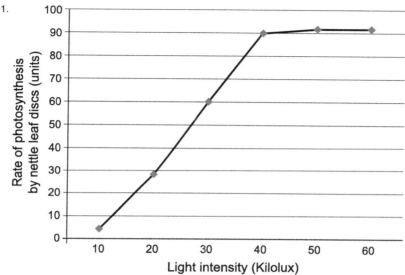

2. Effect: (increase/rise) justification: (More CO_2 for photosynthesis to take place)
   Effect: (stays the same) justification: (temperature or other factor limiting rate of photosynthesis)

Q25:

1. rapid increase at low levels of CO_2 (between 200-400 ppm); then increase slows down between 400-800 ppm; and greater increase from 800-1000 ppm; then levels out above 1000 ppm. (Any three points = 2 marks, two points = 1 mark, one point = 0 marks)
2. 5 kg m^{-2}
3. 100%

Q26: Carbon dioxide is an essential material for photosynthesis; increase in carbon dioxide increases rate of photosynthesis. (Two points = 1 mark, one point = 0 marks)

Q27:

1. Increase in yield from 1 kg m^{-2} at 15°C to 5.5 kg m^{-2} at 25°C. (Quantitative data = 1 mark)
   Explanation: increase in temperature increases rate of enzyme reactions; enzymes are involved in the Calvin Cycle. (Two points = 1 mark, one point = 0 marks)
2. 450%
3. Any from water / mineral ions / named ion / disease / genetic factors of variety of tomato / pollination factors such as insects.
4. The temperature may be too high after 25°C for the enzymes present in the Calvin Cycle to work at their optimum, therefore photosynthesis will decrease.

© HERIOT-WATT UNIVERSITY

Topic 3: Plant and animal breeding

Field trials: Question (page 287)

Q1:

Design feature	Reason for carrying out this procedure
Randomisation of treatment	To eliminate bias when measuring treatment effects.
Number of replicates	To take account of the variability within a sample.
Selection of treatments	To ensure fair comparison.

End of Topic 3 test (page 294)

Q2: Yield, nutritional value, resistance to pests and diseases, physical characteristics suited to rearing and harvesting.

Q3: In inbreeding, selected plants or animals are bred for several generations until the population breeds true to the desired type due to the elimination of **heterozygotes**.

Q4: c) inbreeding depression.

Q5: This prevents/reduces insect attack OR there is less damage to the plant by insects. Photosynthesis is greater/not reduced OR food is available for growth.

Q6: The insect has gained resistance/developed tolerance to the toxin.

Q7: b) the hybrids are heterozygous and therefore not true breeding.

Q8: As sward height increases from 4 cm to 10 cm, milk production increases from 12 kg/day to 18 kg/day; As sward height increases further from 10 cm to 16 cm, milk production remains constant at 18 kg/day; As sward height increases further from 16 cm to 18 cm, milk production decreases from 18 kg/day to 17 kg/day.

Q9: 10 cm

Topic 4: Crop protection

Problems with pesticides: Questions (page 306)

Q1: d) phytoplankton and dolphin

Q2: c) 130

End of Topic 4 test (page 310)

Q3: The crop yield would increase because weeds would be killed. Weeds would no longer compete with crop plants for light, water, minerals. Crop plants will have more resources and grow better.

Q4: They could interbreed and produce weeds which have the gene for glyphosate resistance, therefore weedkiller would be ineffective.

Q5: c) 4050

Q6: b) Pea

Q7: c) Leather jacket

Q8: Rapid growth, short life cycle, high seed output, long-term seed viability.

Q9: a) herbivorous fish and carnivorous fish.

Q10: c) 1.2×10^5

Q11:

- Pesticides can cause problems in the environment because they can **accumulate** within the body of an organism.
- They can also **magnify** along food chains.
- This means each successive organism in the food chain has a **higher** concentration of the chemical in its tissues than the previous organism.

Q12: 80%

Q13: Insecticides kill insects which damage crops/reduce crop yield.

Q14: Insecticides are sprayed onto crops, but rain washes it into rivers. Fish pick up insecticide from water, or from eating microscopic animals which contain it.

Q15: The use of pesticides may result in a population selection pressure producing a **resistant** population.

Q16: d) Chemical, cultural and biological

Topic 5: Animal welfare

Animal welfare: Animal freedoms (page 315)

Q1:

Freedoms for Animals	Example
Freedom from hunger and thirst.	Animals should be able to drink fresh water when they need it.
Freedom from chronic discomfort.	Animals should be kept in a comfortable environment.
Freedom from pain, injury and disease.	Environment should be safe for animals and not cause them injury.
Freedom to express normal behaviour.	Animals should be able to move around freely and mix with other animals in the group.
Freedom from fear and the avoidance of stress whenever possible.	Animals should not be exposed to unnecessary pain.

End of topic 5 test (page 318)

Q2: d) 1, 2 and 3

Q3: a) 1 and 2

Q4:

Type of abnormal behaviour	Example of abnormal behaviour
Stereotype behaviour	Polar bears pacing in a zoo
Misdirected behaviour	Tail biting in pigs
Failure in sexual behaviour	Cheetahs unable to breed in captivity
Altered levels of activity	Hysteria among turkeys

Topic 6: Symbiosis

Extended response question: Mutualism and parasitism (page 328)

Suggested marking scheme

Each line represents a point worth one mark. The concept may be expressed in other words. Words which are bracketed are not essential. Alternative answers are separated by a solidus (/); if both such answers are given, only a single mark is allocated. In checking the answer, the number of the point being allocated a mark should be written on the answer paper. A maximum of six marks can be gained.

A) Mutualism *(maximum of 2 marks)*:

1. Mutualism is a form of symbiosis.
2. Mutualism is a close/intimate/coevolved/long-term relationship.
3. This relationship is one in which both species benefit.
4. Both species have evolved over a long period of time to be dependent upon each other.

B) Parasitism *(maximum of 4 marks)*:

i. Defined as interaction between two species where host is harmed and parasite benefits.
ii. Developed by coevolution/coevolved.
iii. Parasite benefits as it gains energy/nutrients/resources.
iv. Negative to host since resources/energy are lost.
v. Parasites can have limited metabolism.
vi. Often cannot survive outside host/reproduction requires host.
vii. Brief description of one method of transmission such as direct contact or vector.
viii. Example of a parasite including the name of the parasite and its host.

ANSWERS: UNIT 3 TOPIC 6

End of Topic 6 test (page 329)

Q1: An organism which carries disease from one individual to another (1 *mark*) without suffering from the disease (*1 mark*).

Q2:

A) Add fish to the water to eat eggs or larvae/drain wet areas/spray water with insecticide.
B) Add oil or detergent to the water surface to stop pupae breathing and prevent adults emerging.
C) Use insecticides to kill adults/ use mosquito nets or repellent to prevent being bitten.

Q3: d) mutualism.

Q4: b) Light; no food supplied

Q5: Vectors, intermediate (secondary) hosts, direct contact or resistant stages.

Q6:

1. One species benefits and the other is harmed. **Parasitism**
2. Both species in the interaction benefit. **Mutualism**

Q7: Column B

Q8: Row D

Q9:

1. Mutualism
2. *Any one from*:
    - alga is afforded shelter and protection;
    - alga obtains carbon dioxide from host;
    - alga obtains nitrogen compounds from hosts' excretory waste.

Topic 7: Social behaviour

Social hierarchy: Questions (page 337)

Q1:

Contest	Score out of 20 (points)	Winner	Net number of contests won
T v Q	T 17, Q 3	T	14
T v R	T 3, R 17	R	14
P v Q	P 18, Q 2	P	16
Q v R	Q 0, R 20	R	20
Q v S	Q 8, S 12	S	4
R v P	R 13, P 7	R	6
P v T	P 14, T 6	P	8
S v T	S 5, T 15	T	10
R v S	R 19, S 1	R	18
S v P	S 4, P 16	P	12

Q2: Bird Q. *Explanation*: Bird Q was dominated by all other birds

Q3: Bird R. *Explanation:* Bird R dominated all other birds.

Q4: $R > P > T > S > Q$

Social mechanisms for defence: Questions (page 340)

Q5:

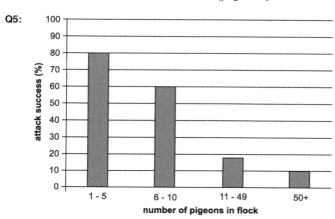

Q6: -87.5 %

Q7: Goshawks are more successful when there are fewer pigeons in the flock; the percentage attack success for 1-5 pigeons in the flock is 80% and decreases as number in flock increases until for 50+ pigeons in the flock the percentage attack success is 10%.

Q8: In a large group it is more likely that a pigeon might see a goshawk; the pigeon can raise the alarm and mass movement might confuse the goshawk.

Altruism and kin selection: Questions (page 342)

Q9: 250 %

Q10: 92.1 %

Q11: In white-fronted bee-eaters, individuals are more likely to help those to whom they are more closely related.

Q12:

- Reciprocal altruism: a behaviour in which an organism acts in a manner that temporarily reduces its fitness while increasing another organism's fitness with the expectation that the other organism will act in a similar manner later on.
- Social hierarchy: society in which some individuals are dominant to others who are submissive to the dominant ones.
- Kin selection: strategies in evolution that favour the reproductive success of an organism's relatives even at the cost of an organism's own survival and reproduction.

Extended response question: Social behaviour (page 348)

Suggested marking scheme

Each line represents a point worth one mark. The concept may be expressed in other words. Words which are bracketed are not essential. Alternative answers are separated by a solidus (/); if both such answers are given, only a single mark is allocated. In checking the answer, the number of the point being allocated a mark should be written on the answer paper. A maximum of ten marks can be gained.

A) Altruism and kin selection (*5 marks*):

1. Altruistic behaviour harms the donor **and** benefits the recipient.
2. Reciprocal altruism involves reversal of roles at a later stage / favour returned or a description of reversed roles.
3. Reciprocal altruism often occurs in social animals/social insects OR mention of the Prisoner's Dilemma.
4. Altruism is (more) common between kin / related individuals / kin selection is altruism between kin.
5. Donor can benefit indirectly (through shared genes).
6. Increased chance of shared / their genes surviving / being passed on (in recipient's offspring).

B) Primate behaviour (*5 marks*):
   i. Primates have a long period of parental care / spend a long time with their parent(s) / look after young for a long time.
   ii. This gives opportunity to learn complex social skills.
   iii. (Social) primates use ritualistic display / appeasement (behaviour) to reduce conflict/aggression / ease tension.
   iv. Any one example of appeasement / alliance forming / ritualistic behaviour e.g. grooming / facial expression / body posture / sexual presentation.
   v. Second example of appeasement / alliance forming / ritualistic behaviour.
   vi. Individuals form alliances which increase social status OR Social hierarchy exists.

End of Topic 7 test (page 349)

Q13: *Any two from*:
- the dogs can tackle large prey animals which they would not be able to tackle individually;
- the dogs will gain more food than they would by foraging alone;
- all members of the social group will share food gained by cooperative hunting.

Q14:
1. 70%
2. The larger the flock the more difficult to focus on/target a single pigeon *or* scattering of flock distracts/ confuses hawk *or* greater chance of hawk being spotted *or* large flock can mob/attack the hawk).
3. More chance/easier to catch prey *or* can catch larger prey *or* all members in the group get a share of the food/prey *or* each member uses less energy/gets more food *or* increase in attack success.

Q15:
- Ritualised threat gesture: social signal used by the leader in a dominance hierarchy to assert authority.
- Subordinate response: social signal used by low-ranking member of a social hierarchy to indicate acceptance of the dominant leader.
- Dominance hierarchy: system of social organisation where the members are graded into a rank order.
- Cooperative hunting: type of foraging behaviour employed by a group of predators resulting in mutual benefits.

Q16: b) less time with its head raised but the group is more likely to see predators.

Q17: d) Predatory gulls have difficulty picking out an individual puffin from a large flock.

Q18: U

Q19: Social hierarchy

Q20: *Any two from*:

- aggression between members becomes ritualised;
- real fighting is kept to a minimum;
- serious injury is normally avoided;
- energy is conserved;
- experienced leadership is guaranteed;
- the most powerful animals are likely to pass their genes onto next generation.

Q21: d) 1, 2 and 3

Q22: The workers are closely related to the queen and her offspring and share similar genes (*1 mark*); by helping the queen to reproduce and in caring for her offspring, the workers are effectively aiding the perpetuation of theirs (and the species) genetic complement through their own altruistic behaviour (*1 mark*).

Q23:

1. $\frac{15}{50} \times 100 = 30\%$
2. $\frac{30}{50} \times 100 = 60\%$
3. Conclusion: Predation of eggs was much higher nearer the solitary nests.
   Reason: By day 7, there were no experimental eggs left near the solitary nest but 20 near the colonial nests.

Q24: Wasps or ants.

Q25: In social insects **few** individuals breed and the offspring are raised by the **workers**. Most of the bees in a colony are **workers** that help to raise close relatives but do not themselves reproduce. This is an example of social **kin selection**.

Q26: Grooming / submissive facial expression / submissive body posture / sexual presentation.

Topic 8: Components of biodiversity

Extended response question: Biodiversity (page 359)

Suggested marking scheme

Each line represents a point worth one mark. The concept may be expressed in other words. Words which are bracketed are not essential. Alternative answers are separated by a solidus (/); if both such answers are given, only a single mark is allocated. In checking the answer, the number of the point being allocated a mark should be written on the answer paper. A maximum of four marks can be gained.

1. Genetic diversity is one component of biodiversity.
2. Genetic diversity comprises the genetic variation of a species.
3. It is represented by the number and frequency of all the alleles in a population.
4. Species diversity is one component of biodiversity.
5. Species diversity comprises the number of different species in an ecosystem (the species richness) and the proportion of each species in the ecosystem (the relative abundance).
6. A community with a dominant species has a lower species diversity than one with the same species richness but no particularly dominant species.
7. Small habitat islands have low species diversity (or converse).
8. The more isolated a habitat island is, the lower the species diversity (or converse).
9. Ecosystem diversity refers to the number of distinct ecosystems within a defined area.

End of Topic 9 test (page 359)

Q1: *Any two from*:
- ecosystem;
- genetic;
- species.

Q2: c) Species richness

Q3: b) Ecosystem

Q4: a) Genetic

Q5: c) Species

Q6: D

Q7: d) The number of different species in an ecosystem and the proportion of each species in the ecosystem.

Q8: Alleles

Topic 9: Threats to biodiversity

The impact of habitat loss: Questions (page 367)

Q1: Habitat fragmentation

Q2: Habitat corridors

Q3: Allow interbreeding between other members of the population which may have different genes OR prevents inbreeding of isolated populations (*1* mark) this provides an increase in variation within the population and prevents extinction of the population if attacked by a lethal pathogen (*1* mark).

Introduced, naturalised and invasive species: Question (page 369)

Q4:

- Introduced - moved by humans either intentionally or accidentally to new geographical locations.
- Invasive - spread and outcompeting native species for space and resources.
- Naturalised - established within wild communities.
- Native - species indigenous to the location.

Extended response question: Introduced species (page 370)

Suggested marking scheme

Each line represents a point worth one mark. The concept may be expressed in other words. Words which are bracketed are not essential. Alternative answers are separated by a solidus (/); if both such answers are given, only a single mark is allocated. In checking the answer, the number of the point being allocated a mark should be written on the answer paper. A maximum of six marks can be gained.

1. Introduced species are those that humans have moved (either intentionally or accidentally) to new geographic locations.
2. Those that become established within wild communities are termed naturalised species.
3. Invasive species are naturalised species that spread rapidly and eliminate native species.
4. Invasive species may well be free of the predators/parasites/pathogens/competitors. (*any two*)
5. ... that limit their population in their native habitat.
6. Introduced species may prey on native species.
7. Introduced species may outcompete native species for resources.
8. Examples of introduced species and their impact on indigenous populations e.g. introduction of the grey squirrel to the UK or cane toad to Australia.

© HERIOT-WATT UNIVERSITY

End of Topic 10 test (page 371)

Q5: c) Smaller

Q6: Habitat corridor

Q7: Invasive species may be free from (*any one*):

- predators;
- parasites;
- pathogens;
- competitors.

OR Invasive species may (*any one*):

- prey on native species;
- outcompete native species for resources.

Q8: Bottleneck

Q9: a), b) and d)

Q10: 8 months

Q11: 33.3%

Q12: In February there is more light for photosynthesis on the forest floor, because leaves have not grown on the trees yet.

Q13: Wide variety of different populations/species present in the wood.

Q14: Drastic reduction in numbers/mass extinction of numbers of squirrels caused by climate change/competition/habitat destruction/population bottleneck.

Q15:

1. Decreased
2. Cheetah, tiger.

Q16: C, the aphid.

Q17: A

Topic 10: End of unit test
End of Unit 3 test (page 376)

Q1: Photosynthesis

Q2: The human population is increasing.

Q3: a) 1 and 2 only

Q4: Line C

Q5:
1. X - NADPH, Y - sugar OR glucose
2. Calvin Cycle / carbon-fixation
3. Rubisco
4. Cellulose

Q6: *Any one from*:
- higher yield;
- higher nutritional value;
- resistance to pests;
- resistance to diseases;
- improved physical characteristics suited to rearing and harvesting;
- ability to grow in a particular environment.

Q7: b) Randomisation of treatment.

Q8: a) Number of replicates.

Q9: c) Selection of treatments.

Q10: Inbreeding

Q11: c) High seed output, and d) rapid growth

Q12: c) biological control.

Q13: *Any one from*:
- toxic to animal species;
- persist in the environment;
- accumulate in food chains;
- magnify in food chains;
- produce resistant populations.

Q14: d) 1, 2, 3 and 4

Q15: Stereotype(s) or stereotypy.

Q16: b) Benefits the parasite and harms the host.

ANSWERS: UNIT 3 TOPIC 10

Q17: Mutualism

Q18: *Any one from*:

- vectors;
- intermediate (secondary) hosts;
- direct contact;
- resistant stages.

Q19: a) 1, 2 and 3 only

Q20: Bees, wasps or ants.

Q21: Kin selection

Q22: *Any one from*:

- grooming;
- sexual presentation;
- facial expression;
- body posture;
- gesture.

Q23: Long period of/extended parental care. OR Look after/stay with young for many years.

Q24: Genetic

Q25: b) species richness.

Q26: b) It will decrease.

Q27: Habitat fragment 1

Q28: Habitat corridors

Q29: Bottleneck effect

Q30: a) Evening primrose

Q31: b) Rhododendron

Q32: c) Pink sorrel

Q33: To make the results more reliable / to reduce the effect of atypical results.

© Heriot-Watt University

Q34: *Any two from:*
- size of leaf disc;
- diameter of leaf disc;
- mass of leaf disc;
- surface area of leaf disc;
- leaf thickness;
- concentration of solution;
- volume of solution;
- temperature of solution;
- size of syringe;
- distance from light source.

Q35: They produced oxygen which made them more buoyant/lighter.

Q36: They can photosynthesise well at low light intensity.

Q37:

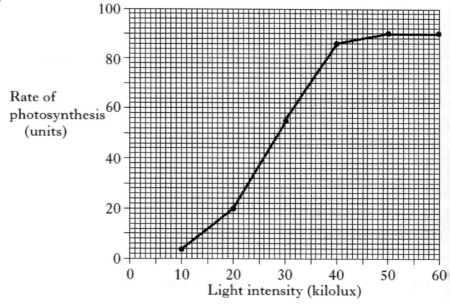

1 mark for correct labels including units.
1 mark for correct scales.
1 mark for plotting the points correctly and connecting them with a ruler.

Q38: 175%

Q39: They have carotenoids / they have more carotenoids.